'Newsam has a keen technical knowledge of "out there", coupled with an entertainer's compulsion to communicate the subject in a fun way. Infectiously enthusiastic, Newsam makes the universe burst into life.'

Engineering and Technology Magazine

'Fantastic ... one of those books that is crammed with technical stuff, but by some miracle of authorship is simultaneously a really good read in which it's just about possible for the non-technical reader to understand enough of what's going on to realise that there's way much more to the Universe than meets the eye.'

The Bay Magazine

*Everything
You Ever
Wanted
to Know
About the
Universe*

Everything You Ever Wanted to Know About the Universe

And our place within it

ANDREW NEWSAM

Elliott&Thompson

First published 2020 by
Elliott and Thompson Limited
2 John Street
London WC1N 2ES
www.eandtbooks.com

This paperback edition first published 2022

ISBN: 978-1-78396-649-3

9 8 7 6 5 4 3 2 1

A catalogue record for this book is available from the
British Library.

Cover design by Jo Walker
Typesetting by Marie Doherty
Printed in the UK by TJ Books Ltd

FSC
www.fsc.org
MIX
Paper from
responsible sources
FSC® C013056

CONTENTS

INTRODUCTION

'How does that work?' It is a question we have all asked ourselves many times, and that curiosity – which goes beyond the necessities of life, and picks and niggles at everything around us – is arguably the spark that gives humanity its special place in the animal kingdom. Out of curiosity has grown science – a powerful set of techniques for answering questions and sharing what we learn. Much of what we have discovered has been practical and aimed at day-to-day problems and challenges. Whether it is biology giving us the knowledge of life that drives medical advances, the insight from chemistry that allows us to create better, more versatile materials, or physicists' discovery of the electricity that powers our modern world, science can claim many advances that have benefited everyone (as well, of course, as a few more worrying 'improvements').

But some areas of science have less obvious practicality, of which astronomy is the most obvious example.

So, what is the point? Why do we spend so much time and effort studying things far beyond the Earth – some so far away that by the time we see them, they probably do not even exist anymore? It is tempting to highlight the unexpected practical benefits, of which there are many. Helium was first discovered by studying the Sun, and it now plays a vital role in many fields as a refrigerant (not just for party balloons and silly voices!); digital cameras and the advances in computing that have revolutionised many of our lives were driven in part by the needs of astronomers; and even esoteric, theoretical physics such as Einstein's General Theory of Relativity is now vital to GPS and SatNav systems, over a hundred years after it was formulated.

But to me, these examples miss the point. Yes, it is always good to find new and valuable ways to use knowledge and discoveries, but the driver is, and must be, pure curiosity – the need to know. The fact that you have picked up this book is a sign that you are also curious and have that desire for discovery, and what better to be curious about than the whole of existence and our place in it? Welcome to astronomy.

WHERE TO START?

Taking an interest in astronomy can be as simple as looking up. The night sky is open to anyone. With the increase in street lighting and pollution it is getting harder and harder for us to see but, nevertheless, on any clear night there is always something to look at, even in the middle of the largest, most over-crowded city.

The first and most obvious thing to look for is the Moon. It is a common sight, but surprisingly few people take the time to really *look* at it. One of the most wonderful things about the Moon is how much it changes when viewed through even the smallest telescope or binoculars. I carry a tiny telescope (about the length of my little finger) with me all the time, and when I spot the Moon, I usually take a moment to look at it in close-up. This is particularly special when the Moon isn't full and the terminator (the dividing line between dark and light) is clear, showing the jagged line of mountains and their shadows.* If you are lucky you can even catch tiny bright spots just into the dark side of the Moon where the rays of the Sun catch the tops of mountains.

* *

*   See plate section.

Also visible even in light-polluted skies are the brighter planets: Mercury, Venus, Mars, Jupiter and Saturn. You have probably seen them many times without realising. Unfortunately, they move around from night to night and month to month, and are often not in the night sky at all, so knowing where to look is not always straightforward.

Venus is usually easy to spot when it is up, though, as it is one of the brightest things in the night sky (second only to the Moon) and is visible either in the west just after sunset, or the east just before sunrise. If you see a very bright star-like object in the fading glow of sunset, and it doesn't move or blink, it is probably Venus (if it moves or blinks, it is probably a plane). You may also notice that it will 'twinkle' rather less than any stars visible at the same time. This is characteristic of the planets,* so can be

. . . .

* Stars appear to twinkle because turbulence in the air above us blurs the light from the stars, rather like the fuzzy view through the heat-haze of a jet exhaust. Planets are also blurred in the same way, but if you look at a planet through a telescope you will see different parts of the planet appear to get rapidly fuzzy or sharp, and the overall effect tends to cancel out (for each blurred bit, there is a sharp part and so on) and the result is that, on average, the 'twinkle' effect is reduced.

a good way to spot them – a star that does not twinkle is probably not a star.

Once you have found a planet, working out which one it is can be tricky. Mars has a distinctive reddish tint, but Saturn and Jupiter both look quite similar (Jupiter is somewhat brighter, but that is only obvious if they are both visible at the same time). To know for certain, you will probably need to use a star map for the right night. Many newspapers publish such things every day, and lots of websites also have them, but they are increasingly being replaced by smartphone 'sky apps'.

These apps are a huge boon to any stargazer. Rather than having to remember constellations, star names, and the slow dance of the planets, they use the location technology within your phone to show you exactly what you are looking at, often overlaying the names and constellation patterns onto the view through the phone camera. With even the simplest, free apps, anyone can become an expert in the night sky in minutes.

Of course, while more can be seen in a bright city than you might expect, the darker the skies are, the more there is to spot. However, you do not always have to go to the middle of nowhere to see interesting things. Even a relatively dark place in a city or town can make a big difference. In such places – parks, back gardens, unlit streets – as your eyes adjust to the dark*, you will begin to see more stars and perhaps the patterns of some recognisable constellations and nebulae: Ursa Major, sometimes called the Plough or the Big Dipper; Polaris, the North Star – always due north; the distinctive 'W' shape of the constellation Cassiopeia; Orion, shaped like a man with broad shoulders; the Pleiades, or Seven Sisters; and in the Southern Hemisphere, the Southern Cross and Centaurus.

But to see the sky in its full glory, with its numerous nebulae, countless stars and the glowing band of the Milky Way, you need a bit more than a dark-adapted eye. You have to find somewhere truly dark and clear. Fortunately, there is an international effort to find and conserve the best dark skies – the International Dark-Sky Association or IDA.† Still growing, the IDA has an ambitious aim to create a complete map of the best places to observe the sky, so wherever you are you can find somewhere reasonably nearby with suitable dark skies (although good

. . . .

* If you need a light while dark adapting, you can use a red torch – red light has much less impact on your dark adaptation. For this reason, many smartphone sky apps have a 'night vision' mode that turns everything red.

† www.darksky.org

weather may be more of a challenge). You will be amazed at the variety and beauty of the heavens.

GETTING TO GRIPS WITH THE BASICS

Of course looking is only a small part of astronomy. The other part – the part that most excites me – is trying to understand what it is you're seeing and make sense of it all. Sometimes this can seem complex or even overwhelming, but don't be put off.

One of the things people struggle with is the vast scale of the universe. Even in our own solar system, the Moon might seem far away but the Sun is more than 400 times further away, and Neptune tens of times further away still. Once we leave the Solar System, the next things we find are the nearest stars – not just a few tens or even hundreds of times further away than Neptune, but around 10,000 times the distance. It can be very hard to get your head around such huge numbers.

Then zoom out even further. This is the Milky Way – our galaxy, full of stars and planets, dust and gas, all orbiting around a huge black hole invisible in the centre, with our entire Solar System just an insignificant, lost speck.

It would be easy to stop here, and for many years this was all we knew – a single galaxy sitting in a universe of nothingness. However, we now know that this is not the end, but just a beginning. Our Milky Way is just one of billions of galaxies that fill the observable universe, some so far away that the light from them can take billions of years to reach us.

This, then, is the universe we live in and which astronomers like myself are lucky enough to study. A vast hierarchy with planets and their moons orbiting around stars, those stars in turn orbiting around the centre of their galaxy, and the galaxies themselves scattered across the probably infinite universe.

It is easy to feel lost and insignificant against this vastness, but we have something that I think puts it all into a different perspective – knowledge. Driven by our curiosity, humanity has looked into that void scattered with points of light and tried to discover what we are looking at and how it works. There are still many unanswered questions to keep that curiosity keen, but we now know a lot about the universe, and how we fit into it. It is out in space that we discover the origins of the atoms that we are made of, and see the vast numbers of other planets scattered throughout the cosmos that might harbour our alien counterparts. From the tiniest sub-atomic particle to the largest galaxies, and even the ebb and flow of space and time, the universe is full of amazing discoveries and fascinating mysteries. And in the rest of this book, I hope you will join me in exploring it.

LOOKING OUT
FROM EARTH

It is a cold, clear, moonless night eight millennia ago. There are no lights – no fires nearby, not even a distant glow on the horizon from the long-set Sun – and above you stretches the astonishing night sky: the twinkling points of stars, the mysterious pearly glow of the Milky Way, the brief streak and flash of a shooting star glimpsed out of the corner of your eye. What are you thinking as you look up?

From our modern standpoint it is perhaps hard to imagine. We have replaced much ignorance with knowledge — we understand the vast distances of space, the complexity of the objects that fill it and have even begun to examine the blackness between the stars. But what we have not lost, I think, is the awe.

Scientists are often accused of 'taking the beauty out of nature', but I cannot agree. No matter how much I learn about the universe, no matter how far advanced telescopes let me explore beyond the limitations of my eyes, I never cease to be amazed by the astonishing beauty, intricacy and scope of the universe. As Keats said, beauty is truth, not ignorance, and the more truth we find, the more we will appreciate the beauty.

Indeed, that beauty is one of the richest joys of being an astronomer. Sometimes the objects we see in the sky are themselves astonishingly artistic — from glowing nebulae to the sweeping majesty of the Milky Way itself, from complex cloud patterns on Jupiter to the rippling curtains of aurorae. But there is also the beauty that comes from discovery itself. I have spent many nights at telescopes, tired, often cold and usually with a headache from the effects of altitude, but it is always worth it for those moments when I see something new. On one occasion it was a patch of the sky that had never been observed in such detail before and which contained thousands of previously unknown galaxies.*

Another time it was a simple, unexpected kink in the trace from a spectrograph. But the feeling is always the same – a wonderful, tight, warm sensation that you are on the edge of finding out something new and adding a tiny new grain to the mountain of 'truth'.

And the more truth we find out about the skies, the more we can understand our own planet and our place in the universe. This has not been easy; astronomy stands slightly outside the other sciences. During the Scientific Revolution and the Age of Enlightenment – while what we now call physics, chemistry and biology were making rapid leaps as increasingly sophisticated experiments were devised – astronomical research still came down to a simple question: can we explain what we see when we look up at the sky at night?

The distinction here is important. The vast majority of science is driven by *experimentation*. Theories and hypotheses are proposed, predictions made and experiments created to test those predictions. From the results,

* See plate section.

the theory can be refined and further tests proposed, better experiments devised and so on. Unfortunately, this approach does not really work for astronomy. Standard methods of experimentation are not usually possible – a chemist can use a Bunsen burner to heat some chemicals, but astronomers cannot do the same to the Sun to see what will happen.* Astronomy, therefore, is not so much experimental as it is about seeing, *observing* – about looking up.

So how have we progressed from our 'primitive' observations of the night sky – peopled with gods, monsters and heroes – to our modern understanding?

A HISTORY OF LOOKING UP

There is plenty about the skies that we can easily see with the naked eye, cycles and patterns that even early humans would have been aware of – the Sun rises each morning and sets each evening, the Moon progresses through its phases and so on. But even though they are easy to see, they are not so easy to understand.

Without understanding, apparently random celestial occurrences might make the world seem an alarming

---

* This is probably a good thing.

place: eclipses that hide the Sun or turn the Moon blood red; comets appearing and disappearing from the skies (often associated with disaster and the death of kings in several civilisations); shooting stars that occasionally come in impressive storms lasting nights at a time. Even the beautiful glowing, shimmering 'curtains' of the aurorae, or Northern Lights, might be cause for concern without understanding how or why they occur. Early civilisations often deified these phenomena that they could not explain; even the Sun, recognised as crucial for life but not understood, was frequently worshipped as a deity.

There are plenty of patterns and cycles that can be easily observed and predicted that might seem simple at first, but the full story is much more complicated.

Take, for example, the Sun. This does indeed appear to rise and set each day. But only twice a year — around the autumn and spring equinoxes* — does the Sun follow a path from due east to due west. The rest of the

. . . .

* The axis that the Earth spins around is tilted a bit compared to its orbit around the Sun, so during the year, the position where the Sun appears to rise and set drifts a bit north or south of due east or west. However, twice a year the tilted axis and orbit line up and the Sun rises due east and sets due west. These two days are the equinoxes.

year the path gradually drifts north or south, reaching its extremes at the two solstices.* This gives us variations in the length of day and night, as well as the seasons – a shorter day means a Sun lower in the sky at midday, and colder weather, with the converse in summer.

Clearly understanding what goes on in the skies is crucial to understanding what's happening on Earth; not only that, the better they can be understood, and predicted, the better life can be. The patterns above us not only provide important clues to the nature of the universe, but also have a huge impact on human activities: the phases of the Moon are crucial for the successful hunter (animals are more active during full moon); predicting the seasons helps a farmer to plan the planting of crops; a thorough knowledge of the stars and their changes during the year makes it possible to navigate, even on ships far from shore; the tides rise and fall as the Moon orbits around the Earth and its gravity 'pulls' at the seas, but the strength of the tide also depends on the phase of the Moon, with the very highest tides and possible flooding at full and new moon; and so on.

. . . .

* The solstices happen when the Sun rises and sets at its maximum distance north or south, before heading back towards the due east/west of the equinoxes.

So it is no surprise that many of the most impressive relics of early humanity are clearly linked to what we could see in the sky. Careful positioning of standing stones in sacred sites like Stonehenge mean they act as calendars, with the Sun rising or setting in perfect alignment with the stones on special days like solstices. Across the Atlantic in Chaco Canyon, another possible calendar has been made of slabs of rock balanced in such a way that, on certain days, they cast a shadow with just a thin 'dagger' of sunlight falling across spiral patterns carved into other rocks.

There are examples of early attempts to observe and explore these patterns found in many parts of the world, although we can only study the societies that had developed written records. Cuneiform tablets from Babylon from before 1500 BCE, together with extensive sets of observations of a myriad of celestial events carefully recorded over centuries, show the first known mathematical approach to astronomical problems and suggest a clear understanding of the way in which the appearance and disappearance of the planets in the night sky repeated regularly. Indian astronomy was also vibrant with sophisticated observatories and detailed calculations. The Mayans and Aztecs developed very accurate calendars and had a concept of time that is in many ways more sophisticated than that of any other society past or present. While the Western world had (and still has) a 'linear' view of time – with past, present and future separate and in that order – the Aztecs had a 'cyclical' view of time, with events coming full circle and repeating, making the concept of past and future much less important. The precision with which some of these early 'calendars' could measure the passing of the seasons is astonishing, with important days such as equinoxes or solstices determined exactly. Such accuracy came from a detailed study of patterns, with measurements taken day after day, year after year.

However, although they had detailed observations, what they still did not have was an understanding of these patterns, or any concept of what was going on beyond our skies. They kept to observing and foretelling rather than explaining.

Nevertheless, there are some surprisingly early examples in history of people beginning to look at the world around them in a more scientific way, searching for answers. One of the most impressive of these early thinkers was Eratosthenes, who made the first attempt at estimating the size of the Earth by looking at the movement of the Sun.

Eratosthenes was born in the ancient Greek city of Cyrene (now part of Libya) in 276 BCE. Like most educated people of his time, Eratosthenes accepted the idea of a round Earth – the flat Earth view had been abandoned hundreds of years before – however, estimating the size of it was a big challenge. Travelling all around the Earth was not an option, so how could it be measured without, say, leaving Egypt? Eratosthenes' solution – like so many brilliant ideas – was actually very simple.

He knew that at noon on the day of the summer solstice, the Sun would shine directly down a well in the city of Swenet (now Aswan), meaning that the Sun had to be directly overhead. Therefore any stick held upright in Swenet at that time would cast no shadow. However, if he held a stick anywhere further north or south – say, Alexandria – because of the curve in the Earth's surface, the Sun would now be at an angle and the stick would cast a shadow. The length of the shadow would depend on what fraction of the Earth is between the stick and Swenet, so he reasoned that if he could then measure the distance to Swenet, he could scale up and get the circumference of the Earth.

The next challenge was to find the distance from Alexandria to Swenet. It is not known exactly how he did this, but there had certainly been many survey trips in the past, and the time taken to travel between the two cities on camel was also well known. It has even been speculated that he paid somebody to walk from one city to the other. Whatever the method, he came up with a circumference of just under 40,000 km – so very close to the true value of 40,075 km.

Of course, we now know that he could not have got a totally accurate measurement since some of his assumptions were not quite right: for example, the Earth is not a perfect sphere but is slightly 'fatter' around the equator, and Swenet is not perfectly due south of Alexandria. Nevertheless the result is very close, and was based on a sound mathematical method – at the time, such an approach based on logic and evidence, rather than mysticism or religious dogma, was the exception rather than the rule. And it is a good example of how valuable our skies can be as a way of understanding more about our planet.

UNDERSTANDING THE SKIES

The more early people observed the skies, the more they wanted a way to explain our surroundings and what was going on above us. No longer satisfied with mystical or religious descriptions of the universe, people started looking for a different kind of explanation.

The ancient Greek and Hellenic philosophers believed in a geocentric model, with the Earth sitting still at the centre of the universe. This is, of course, very sensible. When we stand outside and look at the sky, it is obvious that we are stationary and everything else is moving around us – or, at least, there is certainly no sensation of movement. It would take many centuries of observations and mathematical calculations to prove that this could not be the case.

But at this point, schools of thought were still heavily influenced by Western religious and philosophical orthodoxy, so new observations, calculations and theories often struggled to gain traction if they went against the accepted status quo. For example, the concept of 'celestial harmony', driven largely by Plato and his followers, claimed that while the Earth itself might be corrupted and imperfect, the universe was perfect and unchanging, and the path of any moving objects must be on 'perfect' spheres.

Of course, there were known to be problems with this simple approach, as not everything seemed to have a regular, constant motion. In particular, there are a few 'stars' that buck the trend and do not follow the same routes as their neighbours – indeed, some of them even occasionally reverse direction for a while. These erratic stars fascinated the ancient Greeks, who called them 'wanderers' or Πλανήτες (*planetes*); we now know they are the five planets visible to the naked eye (Mercury, Venus, Mars, Jupiter and Saturn). But it would take several major developments to unseat the theories that had been dominant for so long.

A NEW WAY OF LOOKING

The first of these developments was a – literally – revolutionary publication: Nicolaus Copernicus's *De Revolutionibus*, in which he placed the Sun at the centre of the universe. This provided a much simpler explanation for the confusing movement of the planets, which appeared to change in speed and occasionally even reverse: the Earth and planets simply revolved around the Sun at different speeds. Although started in 1506 and finished in 1530, the theory was not published until Copernicus's death in 1543. This was not a coincidence: such ideas were not just novel but arguably heretical, as they challenged the fundamental nature of the heavens – the idea that the Earth is at the centre of the universe – and therefore challenged the doctrine of the Church.

But these heavens were now standing on shaky ground: scientific advances were increasingly exposing the cracks in the accepted theories. The accuracy of our measurements had improved remarkably, which was one of the motivations behind Copernicus's theory. Of those collecting this new data, the preeminent one was Tycho Brahe.

Brahe collected a vast set of observations of the positions and apparent brightness* of celestial bodies that was unsurpassed in both scope and precision. Although this was before the invention of the telescope (Brahe is considered the last great non-telescopic astronomer), thanks to his great wealth, his instruments were of very high accuracy, and his systematic and meticulous approach matched them. By his death, his catalogue of planetary positions allowed us to study the skies to a level never seen before.

* *

* Since objects look fainter as they get further away, it is important for astronomers to distinguish between the apparent brightness of an object (how bright it looks to us) and intrinsic brightness (how bright it would look if it was moved close to us). Brahe recorded the apparent brightness of stars, but since he did not know how far away any of them were, he could not know their intrinsic brightness.

Tycho Brahe

Born into the Danish nobility in 1546, Brahe had a lifelong passion for mathematics, astronomy and alchemy. He was also famous for his irascibility. While studying at university in Germany, he got into a bitter argument about a mathematical formula with fellow Danish nobleman (and his third cousin) Manderup Parsberg. After several very public disputes, they decided to solve the matter with a duel (not something that is common in mathematical circles now). The duel was carried out in the dark, and both of them survived, but Brahe was wounded on the bridge of his nose and wore a prosthetic metal nose for the rest of his life.

His death was equally theatrical. While attending a banquet in Prague in 1601, he refused to leave the table to ease his bladder, since that would have been a breach of etiquette. Unfortunately, he developed an infected bladder or kidney, and died in considerable pain eleven days later.

It is worth noting that Brahe rejected Copernicus's model, with the Sun at the centre of the universe. His short-lived alternative was a system in which the Sun and Moon orbited the Earth, while the other planets orbited the Sun.

The trouble was that neither model really matched the data. It was left to Brahe's assistant Johannes Kepler to resolve the problem: using mathematical insight, he began to develop a set of 'laws' that were able to describe the motion of the planets.

KEPLER'S LAWS

Kepler's three laws are still used to model and predict orbital motion – not just planets around the Sun, but anywhere where something small orbits something large. Their formulation has been adapted and improved as mathematics has developed, but the core ideas remain the same, which is an astonishing achievement. The three laws can be described non-mathematically as: 1) planets move around the Sun in elliptical orbits, with the Sun at one focus of the ellipse; 2) the speed of a planet varies in relation to its distance from the Sun (i.e. as it gets closer to the Sun, it speeds up); and 3) the length of a planet's year is related by a simple mathematical expression to the size of its orbit.

These laws proved to be far more accurate – and far more useful for prediction – than any of the previous models. Nevertheless, they were still not immediately accepted. The trouble with making progress in astronomy is being

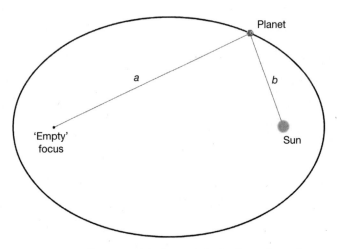

An ellipse. For any orbit, the main body (here the sun) is at one 'focus' with the other empty. The distance from one focus, to the planet, to the other focus (i.e. a+b) is constant, which gives the ellipse it's unique shape.

able to prove a theory. That is still a challenge today, but back then it was even harder. However, a marvellous new invention was to help change all that, one that is still key to our ongoing understanding of the universe today: the telescope. It started with Galileo.

SCIENCE TAKES THE LEAD

In astronomy, Galileo's main legacy is his use of the telescope. Contrary to popular myth, he did not invent the telescope. That honour is usually given to Hans

Galileo

Galileo is often described as the first scientist. This is rather unfair to the many people before him who developed the scientific approach. However, he was probably its most successful promoter at a period when it started to become more widespread.

He was, above all, a passionate believer in experiments. Abstract philosophies did not appeal to him unless they could be tested in some practical way. One of his earliest experiments involved two pendulums. Using his own heartbeat as a 'stopwatch', he changed various aspects of the pendulums – weight, length, length of swing and so on – and found that the time of each sweep varied with the length of the string, but not with the weight of the bob or the length of the swing. This surprising result was the beginning of a revolution in clock-making.

Lippershey who tried, unsuccessfully, to patent a design in the Netherlands in 1608. However, Galileo was one of the first to see the astronomical potential of the invention, and certainly the first to widely publicise his findings.

One of his first discoveries was 'companion stars' near to Jupiter, which he saw moving backwards and forwards from night to night. It was clear to him that these were orbiting Jupiter in the same way that the Moon was thought to orbit the Earth, implying that Jupiter could be another world like the Earth. He also saw that Venus has a full set of phases much like the Moon, something that would have been impossible if the Earth was at the centre of the solar system. Perhaps most importantly, though often forgotten, he also saw that the Moon was marked with craters, mountains and scars, which was not compatible with a perfect heaven.

Of course, these views did not make him popular with everybody. Called to the Inquisition to defend his views and some of his theological writings, he was found 'vehemently suspect of heresy', forced to recant, and spent the rest of his life under house arrest. According to popular legend, immediately after recanting under threat that the Earth moves around the Sun, he was heard to mutter, '*E pur si muove*' (And yet it moves). Whether this is true or not, it certainly fits with the dogmatic character of the man described by his contemporaries, and it may have comforted or perhaps amused him to know that, a mere 350 years later, the Catholic Church would forgive him and apologise for its persecution.

The Scientific Revolution in astronomy was not quite

finished though. The next key development came with the work of one of the most incredible minds in recorded history: Isaac Newton.

GRAVITY

There is almost no branch of physics that does not owe a major advance to Newton, but he is most often linked with gravity. Whether the story of an encounter with an apple in his garden at Woolsthorpe Manor is true or not, his concept of an invisible force of attraction between objects is profoundly important. Not only does it explain why (and how) objects fall on Earth, Newton also said that the force exists between any two bodies in space (e.g. a planet and its moon) and that its strength depends on their mass (i.e. how much they weigh)* and the inverse square of their distance. The 'inverse square of their distance' means, for example, that the force of gravity becomes much weaker the further away the bodies are from each other – if you double the distance, the force of gravity

. . . .

* Weight and mass are slightly different things. On Earth they are essentially the same, but if, for example, you were to take a hammer to the Moon, it will have the same *mass* as on Earth, but will *weigh* less since the gravity is weaker there.

Isaac Newton

Newton's impact on science, and indeed Western thought, is unrivalled. He was a true genius and a polymath – as well as making major advances in mathematics and astronomy, he also invented the milled coin* to inhibit forgers and 'clippers'. He was reclusive and irascible, getting into ferocious and bitter arguments with colleagues and rivals. Even his famous and apparently generous description of his work, 'If I have seen further it is by standing on the sholders [sic] of Giants,' may have been a veiled insulting reference to his rival Robert Hooke, who was slightly built with a spinal curvature from his youth. He was also willing to push himself to his limits. When conducting an experiment on optics, he repeatedly thrust a darning needle down the side of his eyeball and waggled it to distort the lenses in the eye and observe the effect. (Apparently you see 'severall white darke & coloured circles' which disappear when the needle is held still. Personally I'm willing to take his word for it.)

....

*   The set of grooves running around the edge of a coin is called 'milling'. This milling made it much harder to 'clip' the edge of a coin and steal the gold or silver, as the (very precise) grooves had to be replaced.

is four times weaker; if you treble the distance, it is nine times weaker.

The idea of an 'inverse square law of attraction' was not unique to Newton, but he went far further by developing an entirely new branch of mathematics to explore the consequences. Now called 'calculus', this is probably the most important tool developed in the last 2,000 years of mathematics, and it enabled Newton to show that his theory worked with Kepler's laws. We now had a true explanation of the skies, and science had come of age.

BEYOND THE EYE

As already noted, it's hard to test the theories developed in astronomy, so being able to observe more than we can see with the naked eye has been crucial – hence the importance of the telescope both then and now, as we continue to look ever further into the reaches of space.

A telescope is in essence very simple: it allows us to see things that are otherwise too far away or faint. It does this by gathering as much light as possible and bringing it to a focus where it can be examined – either by eye, or using a camera. The earliest telescopes used lenses to gather the light, and this is still the popular image of the telescope, but lenses – especially large lenses, which we need to see further into space – can be unreliable.

Therefore, most modern large telescopes use a mirror to gather the light. These mirrors are not flat (there is little point sending the light from a distant star straight back where it came from) but carefully curved to bring the light to a focus. By using high-quality coatings (usually silver, aluminium or gold) and keeping the mirror very clean, more than 90 per cent of the light can be gathered accurately. This means that telescopes are now built with a main (or 'primary') mirror up to 10 metres across, and construction is underway of 40-metre telescopes which will gather more than a thousand million times as much light as the eye.

One of the reasons modern telescopes need to be bigger and gather more light is to deal with problems of faintness and resolution. As we look deeper and deeper into space, the objects that we're looking for get fainter and fainter – light follows an inverse square law just like gravity (so again, if you double the distance to something it gets four times fainter; treble the distance and it becomes nine times fainter). Even something as bright as the Sun becomes absurdly faint when it is a few billion times further away. Look far enough, and eventually bigger telescopes are needed to gather sufficient light to detect and study such faint objects.

Sensitive cameras can help, but resolution (the sharpness of the image) can be tricky. Bigger lenses or mirrors will produce sharper images, but there is a limit to the resolution that can be achieved, due to the air around us. Light from stars and galaxies travels through the vacuum of space for countless millions of kilometres, in a straight line, but that all changes when it gets to the thin layer of atmosphere surrounding the Earth. When light travels from one medium to a more or less dense one, it changes direction slightly. This is called refraction, and it is very obvious if you look at a straw in a glass of water – the straw seems to bend where it goes into the water. The same happens in the atmosphere, which means that the light gets 'bounced around' as it comes towards us, and the moving and changing pockets of air in the atmosphere cause the stars to appear to 'twinkle'. So, no matter how large your telescope and how good its optics, there is a limit on resolution – on what we're able to see – from inside the atmosphere.

However, some places on Earth are, on average, better than others for observing. Being very high up helps, simply because you are pointing your telescope though less air, and a smooth flow of air above your site will help as well. This makes volcanic mountains on tropical islands ideal, forcing astronomers to spend lots of time in places like Hawaii and the Canary Islands – it can be a tough life.

Even on the best observing sites, however, there will always be limitations. Unless, of course, we can gather the light from the universe before it gets messed about by the atmosphere. In other words, unless we have space telescopes.

ASTRONOMY FROM SPACE

The most famous telescope in space, orbiting the Earth, is probably the Hubble Space Telescope or HST (indeed, most people think it is the only one). The images from the HST are ubiquitous and stunning, and its scientific contributions have been breathtaking. The reasons for this are not hard to see. Although not a particularly large telescope – its primary mirror is just 2.4 metres across and so dwarfed by ground-based telescopes – simply by being above the turbulent atmosphere, it has been able to get extremely high-quality images.

HST didn't have an easy start, though. The idea for an 'extraterrestrial observatory' had been around since the early twentieth century and the success of early, small telescopes like the Orbiting Astrophysics Observatory in the 1960s persuaded NASA that developing and launching a large, orbiting observatory with a telescope containing a mirror 3 metres across should be a key goal. However, the substantial cost for such an ambitious project was too much for the Senate, and eventually a budget was agreed in 1978 for a smaller, 2-metre mirror, as part of a collaboration with the European Space Agency to spread the costs.

All was looking set for launch in October 1986, but the accident that destroyed the Challenger Space Shuttle in January 1986 brought the entire NASA Space Programme to a halt. It wasn't until April 1990 that the shuttle Discovery finally carried the HST to its position in low Earth orbit.

However, Hubble's troubles were far from over. Within weeks of its launch it was clear that something had gone quite badly wrong, as the images appeared out-of-focus and far poorer than expected. It turned out that, somehow, the primary mirror was slightly the wrong shape. After much modelling and design work, a solution was found. Because it was orbiting close to Earth, astronauts were able to return to the telescope and install a fix – a special set of additional mirrors that corrected for the distorted primary mirror and brought the universe back into sharp focus.*

In the decades since, Hubble has proved its worth many times over. There is no branch of astronomy where it has not led to important discoveries, and in many cases it has revolutionised what we know. Whether it has helped us to study the ebb-and-flow of giant storms on Jupiter, see stars forming in vast clouds of gas, or discover the most distant, and oldest, galaxies we know of, it has never failed to surprise and excite. But beyond that, it has also helped to bring astronomy to a wide range of people. The beautiful images created from HST's observations by the project's team of scientists, designers and artists have become ubiquitous and have led to an ongoing renaissance in 'astrophotography'.†

As well as giving sharper, clearer images than is possible from the ground, space telescopes also allow us to see 'invisible light'. We live in a technological world that makes use of many different kinds of invisible light: radio waves, X-rays, microwaves and many more. We now know that all of these are different kinds of electromagnetic

. . . .

* See plate section.

† See plate section.

radiation, but just a couple of hundred years ago, even the concept of other kinds of light was unknown, and it took an astronomer to discover them: Sir William Herschel.

Herschel is another leviathan of astronomy who we will hear a lot about in this book. Discoverer of Uranus, creator of some of the greatest telescopes ever built, patient (even obsessive) cataloguer of the skies, Herschel probably did more for eighteenth- and nineteenth-century astronomy than anyone else in the world. But perhaps his most important contribution to science was almost accidental (as many great discoveries are).

In 1800, Herschel was studying rays of sunlight. Taking a thin beam of light from a hole in a curtain, he used a prism to split it up into all its colours. He then used a set of thermometers to measure the temperature of each colour. The first thing he noticed was that the temperature increased gradually as he moved from violet to red – interesting, but not revolutionary. However, being curious, he also placed a thermometer beyond the red where there was no light and it got hotter still. There was light that the eye could not see! These 'calorific rays', as Herschel named them, are what we now know of as infrared radiation, and they were the first sign that 'visible' light was far from being the whole story.

Since then, many other kinds of 'rays' have been dis-covered, and in 1865, James Clerk Maxwell showed that all these different kinds of ray – from radio waves to X-rays – are different aspects of the same thing: the 'electro-magnetic spectrum'. It is each ray's energy that makes it unique. He also showed that they travel at the same (vast) speed: the speed of light.

From an astronomical point of view, this is potentially fascinating. Since the rays have different energies, they must be made in different ways. Visible light, for example, is automatically produced by hot objects. Anything that is between about 1,000°C and 20,000°C will visibly glow, with the colour becoming less red and more blue/white as the temperature increases. This is what makes stars visible, since their surfaces are typically a few thousand degrees Celsius.

However, at rather lower temperatures, visible light is no longer produced – infrared radiation is emitted instead. Therefore, if we want to see cooler objects than stars, we need infrared telescopes to detect them. Similarly, very hot things (millions of degrees Celsius) produce radiation that goes far beyond the visible, through the ultraviolet and into the X-ray region. Therefore, if we want to study the hottest things that exist (and we do!), then we need to have X-ray or even gamma-ray telescopes. And even objects producing visible light, such as stars, can become invisible if that light is blocked by clouds of interstellar dust, but infrared light passes through most dust clouds.

By the middle of the twentieth century, it was becoming increasingly clear that by limiting our telescopes to the tiny part of the spectrum that we call 'light' we were missing most of the story. However, the atmosphere again threw a spanner in the works. It is not a coincidence that our eyes have evolved to see such a narrow range of light. Most other kinds – from microwaves through to the most energetic gamma rays – are blocked, reflected or absorbed by the atmosphere. So, for all the other kinds of 'light' we have to again get above the atmosphere and put our telescopes in space.

This is difficult, risky and expensive. Early attempts using high-altitude balloons produced some astonishing results, but these 'observatories' were short-lived (usually lasting hours at most) and hard to control, and often the data was lost when the balloon was pulled off course by strong winds, or the return parachute failed and the instrument crashed into the ground. Only with the advent of satellites and orbiting observatories has the full electromagnetic spectrum really been opened for us to explore, with fundamental and astonishing discoveries coming thick and fast – from the detection of thousands of distant planets, to the discovery of enormous black holes in the centres of galaxies.

Even more recently, our view of the universe has got wider still – not just to other kinds of 'light' but to completely different methods of 'looking'. With developments in particle physics and silicon technology, giant particle detectors buried underground allow us to find elusive particles called neutrinos, which are made when stars explode. And as recently as 2016, after almost a century of searching, we were able to 'hear' the universe for the first time. Not normal sound of course – that does not travel through the vacuum of space – but gravitational waves, tiny disturbances in the very fabric of space and time that spread out from the cataclysmic destruction of, say, a black hole like ripples on a giant pond.

Since the early attempts to understand the night sky, through the first use of the telescope by Galileo, via the mathematical breakthroughs of Kepler and Newton, and on to the technical advances of the last few decades, the invisible parts of the universe have opened up to us, and we now have a much better understanding of the universe and our place in it. Of course, we're always trying to find out more, and one of the things that has been most helpful in our quest has been our studies of our nearest star: the Sun.

2

THE SUN AS A STAR

The Sun is important to us. It has made life on Earth possible, providing the energy needed for plants to grow, warmth to stop us freezing, light for us to explore our world. It drives the weather, moving water and air around our planet. Its great mass creates enough gravity to keep the Earth in a stable orbit, which has barely changed for millions of years and so has allowed time for the gradual evolution of life.

But to astronomers it is even more of interest because the Sun is a star, the only one close enough to us to let us really examine and understand it. Since most of what we have learnt about the universe comes from studying the light of stars, and since the Sun is the star we understand best, it is not going too far to say that the Sun is the key to comprehending the rest of the cosmos.

For something so essential to us, it's clearly important that we understand as much about it as we can. And one of the first questions our ancestors grappled with was how the Sun is continuously hot.

WHY IS THE SUN HOT?

The Sun has been around for nearly 5 billion years, and it is likely to last a similar length of time again, and in all that time it has shone without any significant change in brightness or temperature. Not only does that require a phenomenal amount of energy, it is also remarkably stable – something that was impossible to explain for many years. As early as 450 BCE, the Greek philosopher Anaxagoras was teaching that the Sun shone because it was hot (although he thought it was a red hot stone), but as far as we know, he did not try to explain how it got hot or why it did not cool down. For thousands of years, ideas did not progress significantly, but in the nineteenth century, a realisation that the Earth was much older than previously supposed (i.e. many million rather than a few thousand years old) made the question of how the Sun produces its heat even more pertinent and challenging, resulting in many different attempts to explain the Sun's existence – but none of them quite matched all the evidence.

As the nineteenth century drew to a close, the situation was simple but frustrating. The Earth was now known to be at least hundreds of millions of years old, and probably much older, but even with the most generous assumptions, we could not understand how the Sun could possibly shine unchanged for more than a few tens of millions of years. Something was wrong, some fundamental piece of our understanding was missing, and in 1896 Henri Becquerel found the first part of the answer: atoms could break.

THE ATOMIC SUN

It was a tenet of pre-twentieth century science that atoms were the smallest elements of matter, but the work of Becquerel and Marie and Pierre Curie showed that atoms could, sometimes, spontaneously break apart (or 'decay'), creating light and heat in the process.

The idea of 'atomic energy' was seized upon by two scientists: John Joly from Ireland, and Charles Darwin's son George. They both independently proposed that if the Sun were made of radioactive material such as radium, the problem was solved: atomic energy could keep the Sun hot and largely unchanged for billions of years. So the question now changed: what is the Sun made of?

A new technique called spectroscopy allowed scientists to analyse the chemical composition of something from afar, since different atoms create or absorb different colours.* So now the make-up of the Sun could be understood: the atoms in the Sun were similar to some of those on the Earth, with lots of hydrogen and traces of oxygen, sodium, calcium, magnesium and iron, among others. However, there was no sign of the heavy, radioactive elements needed to explain how the Sun could have existed for so long – another theory was floundering.

There was one unexplained oddity, though. In among the atoms there was one line in the middle of the yellow part of the spectrum that did not match any known chemical element. It was initially thought to be sodium, but in 1868, working from his observatory at his home in Hampstead, an astronomer called Norman Lockyer was able to show that this element was not exactly the correct

. . . .

* See plate section.

colour for sodium and must therefore be a completely new element: helium (named after the Greek god of the Sun, Helios).

This new element fit neatly into the second place* on the periodic table, with a mass four times that of hydrogen. However, a more detailed study showed that actually it wasn't quite four times the mass of hydrogen, but very slightly less. Arthur Eddington recognised the significance of this and proposed a new source of energy for the Sun. He suggested that helium was being created in the Sun by the combination, or 'fusion', of hydrogen atoms, and for every helium atom created, there was a tiny bit of mass left over – that mass, converted into energy, was powering the Sun. The sums added up. This theory could easily explain the energy of the Sun, and the reserves of hydrogen could last many billions of years.

It was an excellent proposal, but at the time the knowledge of the atom was not enough to confirm or deny it. However, physics was making amazing progress in the

. . . .

* The periodic table orders the elements according to their atomic number – the number of protons in each atom, which in turn is related to its mass. So hydrogen is first and lightest with one proton, helium second with two protons, then lithium with three protons and so on.

first half of the twentieth century. With the new field of quantum mechanics,* physicists were able to tackle the problem, and by the dawn of the Second World War, the work of many people (in particular George Gamow, Robert Atkinson, Fritz Houtermans, Edward Teller, Carl von Weizsäcker and Hans Bethe) came together and provided a solution: nuclear fusion.†

In suitably hot, dense, high-pressure conditions, light elements can be pushed together with enough force for them to fuse into heavier elements, losing a small amount of mass as energy in the process. Although such hot, dense conditions cannot easily be created on the Earth, the centre of the Sun is ideal, and with abundant fuel from the hydrogen (which makes up more than 90 per cent of the atoms in the Sun), the problem had been solved.

. . . .

* Quantum mechanics is a branch of physics that developed in the early twentieth century that attempts to explain how the very smallest things (even smaller than atoms) work.

† Note that this is different from the nuclear *fission* we find in nuclear power stations. In fission, very heavy elements are broken into smaller, lighter ones, whereas in fusion, lighter elements are joined together into heavier ones. The crossover point is iron: elements that are lighter than iron can be fused to release energy; heavier ones must be broken apart.

INSIDE THE SUN

With continuous advances being made in science, we now have a good understanding of the workings of the Sun, how it will change and evolve, and what that can tell us about other stars.

We will start right in the centre of the Sun where the energy is being produced: the core. Here the weight of the outer parts of the Sun squeeze down and create an incredibly hot, dense environment. With a temperature of around 16 million degrees Celsius, and a pressure in excess of 260 billion times the pressure on the surface of the Earth, conditions are perfect for nuclear fusion of hydrogen. The reactions in the core convert about 4 million tonnes of mass into energy every second, which may sound like a lot, but that's tiny compared to the total mass of the Sun (2×10^{27} tonnes).*

As well as making the Sun hot and bright, this energy has another important consequence: the heat generated produces a large amount of pressure, which pushes outwards and stops the gravity of the Sun from making it collapse in on itself. In other words, it makes the Sun stable. This is the key to the unchanging longevity of most stars.

• •

* 2,000,000,000,000,000,000,000,000,000 tonnes.

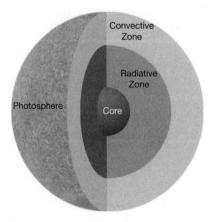

The interior structure of a star like the sun.

About a fifth of the way out from the centre of the Sun, in a region called the radiative zone, the temperature and pressure have dropped enough that fusion is no longer possible. However, it is still quite hot and dense here, and so moving the energy around is not easy. The heat makes its way out from the centre, not in a straight line, but in a complicated, wiggly, random path (sometimes called the Drunkard's Walk). This can take a very long time: from leaving the core to reaching the edge of the radiative zone will typically take nearly 200,000 years.

At the edge of the radiative zone another major change takes place. At this lower density, the gas can start to move around much more easily. Since hotter gas rises, and cooler gas falls, this produces vast currents, where giant pockets of hot gas from the top of the radiative zone rise to the outer edge of the Sun, cool there, and fall back to the edge of the radiative zone, to be heated and rise again. These giant circulating 'convection cells' are similar to the weather patterns on the Earth, but on a far larger scale, and they give this upper layer of the Sun its name: the convection zone.

At the top of the convection zone, the gas is now sufficiently low density for it to become transparent and we can (finally) directly see the effect of all this energy generation and movement. This upper area is called the photosphere, and it is around 5,500°C. It is quite dangerous to look directly at the Sun, but with the right kind of telescope it is possible to see the surface in detail. The most obvious aspect is the overall glow: the Sun is very bright. However, there are also patterns and features. The entire surface is mottled and granular, showing the tops of the giant convection cells, and scattered randomly around are other features: dark 'sunspots'.

The spotty Sun and solar activity

Sunspots have been observed for a very long time. The earliest surviving written record was made in 800 BC by Chinese court astronomers; in 1128, the English monk John of Worcester made the first known drawing of one;

and there are detailed records of about 400 years' worth of sightings since the invention of the telescope. One thing of note from these records is that the number of sunspots drops to nearly zero every eleven years, before climbing up to a peak and dropping down again eleven years later. The peak can sometimes be very high – dozens of sunspots every day – or it can be so low as to be almost zero. We are currently in a high period of sunspot activity, but in the late seventeenth century almost no sunspots were seen for nearly fifty years. This is called the Maunder Minimum and corresponded, probably coincidentally, with a period of very cold winters. One thing that is not coincidental is that when there are lots of sunspots there are also lots of bright aurorae (or Northern Lights) visible here on Earth because both are caused by changes in the magnetic field of the Sun.

Like the Earth, the Sun has a strong magnetic field (indeed, the Sun's is much stronger). It also has an equator and two poles. However, unlike the Earth, the Sun is largely fluid and does not all spin at the same rate: the solar equator spins faster than the regions near the poles. This 'twists' the magnetic field, distorting it and making it very different from the simple field of the Earth. The Sun's magnetic field produces loops and coils that poke up and down through the photosphere, creating sunspots wherever they emerge. It also leads to very large explosions on and above the surface of the Sun. The debris from these explosions is a kind of electrically charged gas, called a plasma, that can be flung away from the Sun at high speed. This is known as a coronal mass ejection (or CME). If the plasma passes by the Earth, it interacts with our own magnetic field, squeezing and distorting it, and producing aurorae and many other effects that together have become known as space weather.

Space weather

Although the aurorae produced by passing CMEs have been seen for as long as animals have had eyes, in recent decades some other, less desirable consequences have come to light. The infrastructure of the modern world (satellites, electricity grids, radio communications, etc.) can also be disturbed or damaged by severe solar activity. Stormy space weather makes the Earth's atmosphere change shape, and satellites can have their orbits changed, or even lowered into the upper atmosphere where they burn up. The sensitive instruments on satellites are also at risk of damage when bombarded by particles, and communications with the ground can be interrupted for quite long periods

– a particular problem for satnav satellites. Large electric currents can be induced in national grid systems, causing blackouts and extensive damage; mobile phone signals can be disrupted; even the internet could be broken. It is only a matter of time before a major solar storm causes significant damage, so the problem of space weather is becoming an important one for governments and insurance companies.

Of course, stopping the Sun from being active is not a realistic (or desirable) option. However, the damage caused by a major CME could probably be avoided if there was enough warning: the solution is not prevention of solar activity, but prediction. As a result, space weather observatory satellites are being designed, built and launched, and some of the massive computational power that is used to model terrestrial weather and climate is being used to explore, understand and, hopefully, predict the weather beyond our atmosphere.

With a better understanding of how space storms batter the Earth, energy companies can design more robust infrastructure; satellite operators can use the improved forecasts to temporarily shut down sensitive systems and prevent damage; the next generation of satnav systems can incorporate better ways of coping with interference from the Sun; and even flight schedules could be altered to minimise the risk of increased radiation exposure on high-altitude flights during very stormy conditions.

From being a niche area of astronomy, the study of solar activity and space weather has transformed in recent years into a major public service, with national organisations such as the Met Office in the UK setting up space weather prediction and modelling units, and providing daily forecasts and alerts to everyone from major communications firms to hopeful aurora spotters.

OUTSIDE THE SUN

The photosphere is the visible extent of the Sun, but it does not actually end there. During a solar eclipse, with the glare of the Sun itself blocked out, a fainter halo can be seen around it. This is the chromosphere: a reddish, irregular layer around 1,600 km deep. It is much less dense than the photosphere, and cooler, which is why it is hard to see against the glare of the Sun, but at the top of the chromosphere something odd happens. Instead of falling further, the temperature begins to rise, rapidly climbing to around 1,000,000°C to create another region we call the corona. This is still not fully understood – in spite of it being so close to us, the Sun still has a few secrets.

The final 'layer' of the Sun is by far the largest. The strong, twisted and distorted magnetic field of the Sun does not stop close to the surface, but continues far into the solar system and out beyond the most distant planets. The magnetic field carries with it some of the plasma from CMEs and other gases blown from the surface of the Sun, and as these gases flow outwards they form the solar wind. The wind is so far-reaching that it arguably forms the boundary of the solar system: outside the wind, the Sun is just another star.

In the last century, our understanding of the Sun has radically improved. Knowledge of the atom has told us how the Sun could produce its prodigious light and heat for so long, and the use of sophisticated satellite telescopes and advanced physics have let us build detailed models of the Sun from the inner core right out to the furthest extent of the solar wind. These are important advances in their own right, but it is as the basis of our knowledge of stars that they form the bedrock of astronomy.

However, before we move out into the reaches of interstellar space, we should not forget the other parts of the solar system: the planets, comets and asteroids that also have a lot to tell us.

3

THE SOLAR SYSTEM

Compared to the deep reaches of space, we know quite a lot about the objects we share our solar system with; we have even physically explored several of them. That makes sense, as they're the most accessible objects available to study – the gaps between us and our neighbouring planets are huge, of course, but they're nothing compared to the vast distances to the stars and planets beyond our solar system. It also makes sense that we want to know more about our immediate neighbourhood. Do the orbiting asteroids pose a threat to us here on Earth? Might life exist somewhere out there? Or would one of these planets

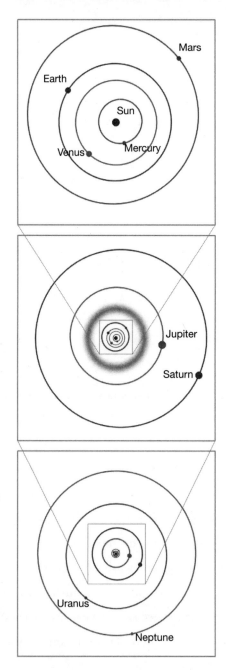

The orbits of the planet to scale.

do as a second home? As our population continues to grow, our planet changes and our resources dwindle, so perhaps we need a back-up plan. As Stephen Hawking pointed out in 2010, 'The human race shouldn't have all its eggs in one basket, or on one planet. Let's hope we can avoid dropping the basket until we have spread the load.'

So, increasing our knowledge is an important task and – as well as studying them with telescopes from Earth – there are many missions to visit and explore the planets, moons and asteroids that make up our solar system. These objects are divided into three broad groups: the rocky terrestrial planets, the giant planets and all the small leftover objects usually just referred to as minor bodies. We will start our tour close to home.

THE TERRESTRIAL PLANETS

The terrestrial planets – Mercury, Venus, Earth and Mars – are found in the inner part of the solar system and they are small planets with rocky, solid surfaces.

Mercury, the Winged Messenger

Visually, Mercury is rather like the Moon: grey, barren, cratered and without any visible atmosphere. However, being close to the Sun, it is much hotter. Despite the many science-fiction stories describing Mercury as a strange twilight land balanced between the constant burning heat of the nearby Sun and the permanent icy cold of the night-side of the planet, 1965 radar observations of Mercury showed that it spins one and a half times every orbit (or exactly three times for every two orbits), so all sides do see both day and night.

Although Mercury appears to have no atmosphere, it does in fact have a very tenuous one.* It is likely that this atmosphere is changing all the time as Mercury is blasted by the solar wind, which strips it away, creating a very faint comet-like tail. The gases are replenished by slower-moving particles in the wind or by gas gradually escaping from surface rocks.

Venus, the Bringer of Peace

Mythologically, Venus is usually associated with beauty, love, peace and prosperity, but the conditions on the surface are closer to a medieval hell. Permanently enshrouded in cloud, we still know very little about Venus's surface,

. . . .

* The atmospheric pressure on Mercury is 10^{-14} bar: 0.000,000,000,000,01 times the pressure on the surface of the Earth.

but we do know that its atmosphere is extremely violent and its surface is the hottest of any planet (nearly 500°C) – even hotter than Mercury, which is much closer to the Sun. This extreme heat is caused by a runaway greenhouse effect.

The greenhouse effect is the main cause of climate change on the Earth. Without an atmosphere, a planet will absorb a certain amount of energy from sunlight and then radiate it away again. However, certain gases in the atmosphere (such as carbon dioxide and water) will send some of the radiation back to the surface of the planet, heating it further.

On the Earth, this is leading to a gradual global increase in temperature as the amount of greenhouse gases in the atmosphere increases, but on Venus things have gone one stage further – a warning sign for us, perhaps. In the distant past, Venus is likely to have had a large ocean. As the temperature increased, the water evaporated, putting more water vapour into the atmosphere, which in turn increased the greenhouse effect, leading to a higher temperature. As the temperature continued to increase, the surface rocks were heated to the point that they began to release stored carbon dioxide, intensifying the greenhouse effect, and again increasing the temperature. The cycle only stopped when there was almost no more gas in the surface rocks to be released, by which time the surface temperature was hot enough to melt lead.

The high temperatures have also led to other nasty atmospheric effects. When it rains on Venus, what falls is not water but sulphuric acid, and the winds at the surface are several hundred kilometres per hour. It is not surprising, therefore, that much of Venus remains a mystery. While probes sent to the Martian surface have operated for many years, the longest-lived surface probe of Venus was the Soviet Venera 13, which landed in 1982 and survived just two hours – still longer than its 32-minute expected lifetime – in the 457°C heat and around 80 times greater pressure than on Earth that it found.

Earth

The Earth is not of much interest to an astronomer, as we have studied it in detail and in situ for millennia, but there is plenty that might be of interest to an alien observer. Our planet has many unique features compared to the others – in particular, our atmosphere is moderately thick and rich in water vapour. Looking closer, a visitor would see large areas covered in oceans, but between them are not just the greys, oranges and yellows of rock and dust, as seen on the other planets, but large areas that are green and clearly not any kind of rock – something of a mystery. Studying the atmosphere would reveal other surprises, in particular the large amount of free oxygen. Since oxygen is a very nasty, reactive gas, it cannot remain free for long but will tend to oxidise almost anything it comes in contact with, so something must be continually replenishing the gas. In other words, the atmosphere of the Earth is out of equilibrium – one of the key signs of possible life. Anyone else looking for a viable home planet would be very keen to visit!

One final anomaly is Earth's satellite, the Moon, which is much bigger than any other moon of any terrestrial planet. Mercury and Venus have no moons at all, and Mars's two moons are much smaller (the largest, Phobos, is only 22 km across compared to over 3,000 km for the Moon).

The Moon

Like Mercury, the Moon has almost no atmosphere and is rocky and heavily cratered. Unlike Mercury, however, it has become fully 'tidally locked', meaning it has the same face pointing towards the Earth all the time. That face, as well as having mountains and craters, also has large, dark, lowland areas called maria, which have far fewer craters than the highlands.

Craters on the Moon are not volcanic, but are formed by the high-speed impact of asteroids. Most of the cratering happened early in the history of the solar system, when there was a lot of debris left over from the formation of the planets flying around. The asteroids that formed the craters were just as likely to hit any part of the Moon's surface, so the cratering should be even — but it isn't. This means that the uncratered areas must have been covered by a new surface since then. This tells us that although the Moon is now solid right through, in the past it must have had a molten rock interior, just like the Earth, which at some point leaked out, filling the lowland areas and quickly cooling and solidifying to leave flat, darker, denser rock that covered the older craters.

This idea was confirmed by observations of the far side of the Moon by the Apollo missions in the 1960s, which showed much more even cratering. This must mean that the liquid core of the Moon was slightly off-centre and so only leaked out on one side — the side which now faces us. The maria and offset core are denser than the older rock, and so the near side of the Moon is slightly heavier, and this is the side that is tidally locked to the Earth — everything fits.

So why is something so big orbiting the Earth when the other terrestrial planets have nothing comparable? The most likely reason is probably the Giant Impact Hypothesis (sometimes called the Big Splash model). The very early solar system was a busy place, and it is suggested that there was another young planet, about the same size as Mars, called Theia with a very similar orbit to the Earth, and it was inevitable that they would eventually collide. A head-on collision would have destroyed both planets, but since they shared an orbit, a glancing blow was much more likely, causing the two protoplanets to merge. This would still have had a dramatic effect on the young Earth, changing its orbit and spin (giving it a day as short as five hours) and also leaving a lot of debris which would have formed a ring around the new, bigger Earth. This ring of rocky debris would have been gradually pulled together by gravity to form a new object: the Moon. This new Moon would have been much closer to the Earth then, and orbiting much faster, but over time the tides pulling between the Moon and the Earth would have slowed the Earth's spin, while making the Moon spiral very slowly outwards, giving us the 24-hour day and one-month lunar orbit that we have today.

This is a very dramatic model, and is far from proven, but there is a lot of supporting evidence: for example, the chemical composition of the Moon and the surface rocks of the Earth are very similar, and different from any other objects in the solar system. It's not settled yet though; the formation of the Moon is still an open question, which we can perhaps try to solve when humans again walk on its surface.

Hopefully that will not be too far in the future. It is odd to think that the climax of the Space Race, when Neil Armstrong became the first human to walk on the Moon in 1969, happened before the majority of people alive today were even born (I missed it by just a couple of months). Since then, the massive cost and high risks of lunar flights have kept humanity to within a hair's breadth of the Earth.* However, things are changing and several countries have plans to resume human exploration of our closest neighbour. Recent discoveries of ice in the permanent shadows of some craters are particularly exciting as the ready supply of water from the ice would mean that it might be possible to set up bases for longer stays, rather than just occasional short visits. This is important as it would also provide a stepping stone to other parts of the solar system. It is far easier to launch missions from

. . . .

* The International Space Station orbits 400 km above the surface of the Earth, while the Moon is nearly 400,000 km away.

the lower-gravity Moon than the Earth, and so journeys to Mars, asteroids or other bodies even further away might become easier.

Of course, this is all a long way in the future, but the American, European and Chinese space agencies, and even India's fledgling space programme, are all exploring the Moon with orbiting satellites and robotic probes, with clear plans to get people back there as soon as possible. (China, for example, has set a return date of 2036, although without publishing a detailed plan of how they might manage that.)

I may not have been born when the first man walked on the Moon, but I hope to live to see the first woman.

Mars, the Bringer of War

Glowing an almost bloody red, Mars is traditionally associated with war and death. However, we now know that it is one of the calmest of the terrestrial planets, with a thin but not negligible atmosphere and only gradual seasonal changes. It is the most studied planet after the Earth, with many orbiting probes, landers and robotic 'rovers' exploring it over the last few decades.

These missions have taught us much about our neighbour planet. The red colour comes from rust – iron oxide – which means that Mars was probably wetter and had a thicker atmosphere in the past. The oxygen that creates the rust is most likely to have come from rain, although carbon dioxide being broken up by the light from the Sun is also a possible candidate. And there is a lot more evidence of past water on Mars, including dry riverbeds, layers of sedimentary rock laid down on ancient seabeds, and even some signs that liquid water still occasionally spills out from underground reservoirs.

Water cannot survive on the surface of Mars now, however – it is too cold. The thin atmosphere is also incapable of supporting much water vapour, so rain is at best very rare and at worst probably impossible. However, even hints of past water are important, since they hold the possibility that Mars may once have supported life. Indeed, if the water has not all evaporated, but is stored underground in salty aquifers or vast ice beds, then perhaps life could survive on the surface – not new Martian life, but us. Of all the planets in our night sky, Mars is the most hospitable to us. If we ever want to move, it is by far the best choice.

Of course, before we can think about migrating to Mars, we need to work out how to live there. It may be relatively hospitable compared to the other planets, but the conditions would still kill a person very quickly. There is no breathable air, no water, no food, it is very cold (averaging about −60°C, compared to a balmy 20°C on Earth) and the thin atmosphere provides little protection from the radiation from space. An unprotected human would suffocate, dehydrate, starve, freeze and die of radiation sickness on Mars.

While it is possible to overcome all of these problems for a short mission – building an air-tight and heated shelter, taking all the food, water and air needed, and so on – the risks would still be enormous. And in the longer term, proper colonisation cannot take place if people are still reliant on a constant traffic of supplies from Earth. Current Martian exploration by satellites and robots, therefore, is concentrating on learning as much as possible about the resources available on the planet to any human missions. Is the water that we think is hidden underground accessible? Can breathable air be generated by a reliable process from the rocks and minerals on Mars? Is it possible to grow food in Martian soil (with some help from Earth bacteria)? Are there caves suitable for living in that could be sealed and used to protect astronauts from radiation? And so on, along with a thousand other unknowns and tests.

Of course, there are also other issues that are far less technical but perhaps more challenging. Foremost is, do we have the right to think about colonising another planet?

At the moment it seems that there is no life currently on Mars, but that could be wrong. What if we brought diseases with us that wiped out what little microscopic Martian life there is? Some people argue that even having the chance of an escape route from Earth makes us less likely to look after our home planet properly. These are not easy problems to solve and while they are not urgent, the rapid progression of technology and our constantly improving understanding of Mars means that it will not be too long before we have to address them properly. Astronomers can provide the facts, engineers can show us what our options are, but the ethical issues are for everybody to think about.

THE GIANT PLANETS

The giant planets – Jupiter, Saturn, Uranus and Neptune – are well named: the Earth could fit inside Jupiter well over 1,000 times. However, it is not their size that most clearly distinguishes them from the terrestrial planets, but their composition. The giant planets contain far fewer of the heavier elements; they are balls of gas with no solid surfaces. Their colours and appearance are due to the mix of gases, and giant weather systems in their upper layers.

Jupiter, the Bringer of Jollity

Jupiter is the giant's giant, containing more than twice the mass of all the other planets combined. Its 'weather' is on a similar scale, with wind speeds of over 300 km per hour being common, and vast storms that survive and swirl for extraordinarily long times: the famous Great Red Spot is a single storm larger than the entire Earth that was first spotted by Galileo over 400 years ago, and it is still going strong.*

However, the most impressive aspect of Jupiter is not actually visible: its magnetic field, which is around 20,000 times as strong as Earth's and millions of kilometres across. If it were visible to our eyes, it would loom larger than the Moon in our night skies, in spite of being over a thousand times further away. It is caused by huge electrical currents circulating deep inside Jupiter, which tells us that the interior of the planet is probably made up of a very strange substance called 'liquid metallic hydrogen' – a bizarre form of the simplest chemical element that only exists under vast pressures and temperatures.

Most of our exploration of the Jovian system has concentrated on its moons. Jupiter has many moons (over sixty at the last count, but there are probably more still to find),

* See plate section.

the majority of which are small, rocky and not particularly interesting. However, four are significantly larger, with their own unique features and mysteries. These Galilean moons (so named because they were discovered by Galileo) can be easily spotted through binoculars as small dots lined up either side of Jupiter, but exploration by the Pioneer, Voyager and Galileo space probes have shown that they are in fact unique, varied and fascinating small worlds.

Io

Io is the closest moon to Jupiter and suffers from very, very strong tidal effects from the massive planet. Where the tides on Earth move water around, the tides on Io are enough to squeeze and stretch the very rock itself. This heats up the interior of the moon, causing continual volcanoes that cover the surface with sulphurous debris and make it look rather like a pizza; Io is by far the most volcanically active body we know of. The gravity of Io itself is not high – less than a fifth of the surface gravity of Earth – so a lot of the hot gases ejected by Io's volcanoes escape into space, where they interact with Jupiter's magnetic field and create vast electrical currents through space – something that spacecraft designers have to be careful of when designing their probes.

Europa

Next out from Jupiter is Europa. This is a rocky body that is in many ways similar to the Moon, except that it has a complete coat of ice. If the surface of the ice were old, we would expect to see craters from meteorite impacts, but there are none (in fact, Europa is the smoothest body in the solar system), showing that the surface is young and being constantly renewed. This raises the exciting possibility that the tidal effects of Jupiter, while weaker than those that torment Io, are enough to heat up the interior of Europa and keep a 'sea' of liquid water under the coating of ice. This makes Europa one of the best candidates in the solar system for extraterrestrial life similar to our own. Because of this, it is one of the main targets for the next big mission to Jupiter. The Jupiter Icy Moons Explorer (or JUICE) is a European Space Agency mission planned for launch in 2022 and which should reach Jupiter's moons by about 2033. With a complicated set of 'flybys', it will use a huge variety of instruments to find out a lot more about Europa and the other moons. However, its primary target is Ganymede.

Ganymede

Ganymede is the largest moon in the solar system, and it is roughly half rock and half water ice. The rock and ice formed layers as the moon formed, and the core of Ganymede is likely to be a giant internal 'ocean': a mixture of water and liquid iron which may contain more water than all the seas of Earth combined. However, since it is further from Jupiter than Europa, the tidal forces are weaker and so the ice covering it is much thicker. Therefore, in spite of its exciting potential for liquid water and hence life, we know even less about it than Europa. JUICE will hopefully change this, with its plans for a full orbital reconnaissance of Ganymede, including taking radar measurements of the thickness of the ice and the quantity of water underneath it.

Callisto

Liquid water may also be a subterranean feature of Callisto, through probably not as much as on Ganymede, or as suitable for life as on Europa. Callisto is an odd object since it is large but low density: it is almost exactly the same size as Mercury, but only about a third of its mass. Perhaps Callisto's most important feature is that it is far enough from Jupiter that the streams of electrically charged particles that fly around the planet's inner magnetic field are far less intense. This is important if we are thinking of sending people to the Jovian system, as these particles are a very dangerous form of radiation and quite difficult to shield against for long periods. This makes the surface of Callisto a much safer place to be than the other moons; if (or when) we start to explore the Jovian moons in person, it would make an ideal base.

Saturn, the Bringer of Old Age

Next out from the Sun comes Saturn. Saturn was the first thing I ever looked at through a telescope. The view of it apparently hanging in the darkness with its elegant, perfect rings is one that I will never forget, and it still enthrals me every time I see it. I am not alone in this: the beauty and mystery of the planet means that it is regularly chosen as a favourite astronomical object, and the gradual change in the 'tilt' of the rings towards the Earth is one that many people follow month after month.

As a planet, Saturn is quite like Jupiter. It has a similar chemical composition, and so a similar appearance, although, being further from the Sun, it is cooler and has less dramatic weather. It is also quite low density: it is the only planet that would float on water (assuming you could find a bathtub large enough). However, the most fascinating thing about Saturn is its extraordinary rings.

When Galileo first saw the rings of Saturn, he thought that they were two companion moons. He wrote, 'Saturn is not alone, but is composed of three, which almost touch one another and never move nor change with respect to one another. They are arranged in a line [...] and the middle one (Saturn itself) is about three times the size of the lateral ones.' After further observations he described Saturn as having 'ears', and was later surprised to discover in 1613 that they had gone, only to reappear a year later.

It took more than fifty years before Christiaan Huygens, using a more powerful telescope, suggested that the 'ears' were in fact thin rings surrounding the planet. As Saturn orbited the Sun, our view of the rings tilted up and down, making them seem to disappear when they were exactly edge on. Not long afterwards, in 1675, Giovanni Cassini saw that there were in fact multiple nested rings with small gaps between them, and it wasn't until 1859 that James Clerk Maxwell showed that the rings were not solid, but made of vast numbers of rocks orbiting together. This becomes clear when we are lucky enough to see a bright star pass behind the rings – the star continues to shine almost undimmed, with just a slight 'flickering' as individual rocks briefly eclipse it. The rings were formed from debris from the early formation of Saturn, and in fact all of the large planets have similar rings, but only Saturn's contains sufficient material to be visible from the Earth.

At first it seems impossible that Saturn's rings stay so clear, crisp and stable, but there are a few larger rocks or small moons that 'shepherd' the ring material and prevent it from drifting too far outwards or inwards. These shepherd moons also cause the small gaps between the rings; in fact, many of Saturn's small moons have been discovered by looking for ring gaps and using them to predict where a moon should be.

Titan

The final surprise that Saturn has for us is its largest moon, Titan, which is second only to Ganymede in size, and slightly larger than Mercury. Unlike any other moon, it has an amazingly thick, dense atmosphere. The other moons have no significant atmospheres since their gravity is not enough to retain the gases: they simply drift off into space. The same must happen with Titan, so the atmosphere must be being continually replenished – probably by 'outgassing' from rocks or sporadic bursts of cold 'volcanoes' – giving rise to a density higher than that of the air around us on Earth. Although we could not breathe it, Titan's atmosphere is surprisingly like that of the Earth

– being rich in nitrogen – and Titan is the only other place in the solar system where large areas of liquid have been found on the surface (though the low temperature means it cannot be water, but liquid methane).

Unfortunately, because it is so far from the Earth, Titan is not as well explored as we would like. Most of what we know about Titan comes from the ground-breaking Cassini mission which, as well as examining Saturn, Titan and other moons from space, also dropped a robotic lander, called Huygens, onto the surface of Titan. It is easy to underestimate the difficulty of this. Because of the great distance to Saturn and the time it takes for radio signals to travel there, no sort of real-time control over the descent was possible, so Huygens was entirely on its own. The uncertainties it faced were daunting: the wind speeds were unknown but were likely to be high; the landing site was also unknown, so the probe had to be able to land on liquid, dust or jagged rocks and still keep operating; and all the time it had to collect, analyse and transmit data on its surroundings. Despite all these challenges and unknowns, on 14 January 2005, after a controlled descent lasting two and a half hours during which data was continually gathered, Huygens successfully landed on the dried-up bed of a methane sea and continued to gather and transmit data on its surroundings for over an hour: far longer than the expected lifespan of three minutes.

The results of the Cassini–Huygens mission have dramatically improved our knowledge of Titan, its atmosphere, surface and weather, but as with all steps forward, just as many questions have been raised as answered. Titan does indeed have liquid on its surface, and the composition of the atmosphere is rich and complex. But the origin of much of that gas is still a mystery, and the most important question is still unanswered: is there, or has there ever been, life on Titan? Future missions may help to answer these questions. There are plans for more landers and perhaps even an autonomous submarine to explore the moon's methane seas, but such projects take time to develop and launch, so for now Titan will remain one of the great mysteries of the solar system.

Uranus, the Magician

Uranus has a special place in the history of astronomy, as it was the first planet ever to be 'discovered'. The planets closer to us are all visible to the naked eye and so have been known about (if not understood) since pre-history, but Uranus cannot be seen without a telescope. The honour of the discovery went to the same man who extended our understanding of light into the infrared: Sir William Herschel.

William Herschel

William Herschel was born in Hanover in 1738 and chris-tened Friedrich Wilhelm Herschel. His father was an army musician and the young Wilhelm also had a talent for music, joining the band of the Hanoverian Guards as an oboist at the age of fourteen. He also played the violin and the harp-sichord and became a very accomplished organist. In 1757, he travelled to England, where he adopted the anglicised ver-sion of his name, William. Herschel's reputation grew and he was able to secure the prestigious job of organist in Halifax Parish Church, where a new organ had been built by the Swiss organ-maker John Snetzler. Herschel astonished the appointment committee with the rich, full tones he was able to draw out of the organ, impressing even Snetzler himself, who declared, 'I will love this man for he gives my pipes room for to speak.' It was only later that Herschel revealed his secret: lead weights placed on the lower keys to augment the harmony.

Herschel's other great passion was natural philosophy. Fascinated by the mysteries of nature since his youth, he soon became a leading light in the intellectual life of Bath, in particular with his knowledge and curiosity about the heav-ens. A very practical man, he began to design and build tel-escopes for himself, starting with a 2.1-metre long refracting telescope. Not long after completing it, on 13 March 1781, he saw an object that at first he thought was a blueish star in the constellation of Taurus. However, his telescope was of far better quality than any other at the time, and with careful observation he saw that this 'star' wandered through the skies just like the planets. He immediately reported his findings to the Royal Society, who confirmed them: the first discovery of a new planet.

Herschel proposed the name Georgium Sidus for the new planet, in honour of George III. The name was not popular outside Britain, however, and it was Johann Bode's suggestion of Uranus that eventually stuck. Herschel's proposed name was not without benefits, however, as he was awarded a pension by the king which gave him the security to devote himself full-time to astronomy. Together with his sister, Caroline, and later his son, John, he built the finest and largest telescopes of his generation and made many discoveries that remain important to this day. In particular, he was one of the first astronomers to realise that detailed catalogues of the sky were needed, and over the course of many years he worked every clear night to do a systematic 'sweep' across regions of the sky, with Caroline taking careful note of every nebulous object that he saw. This was a mammoth undertaking and it was only completed after William's death by his son, John; their catalogue formed the basis for the New General Catalogue, which is still used to this day (many objects are known by their 'NGC number').

Uranus looks very different to Jupiter and Saturn. The reds, oranges and yellows caused by ammonia and similar chemicals in the upper regions of the nearer giants, give way to blue-green on Uranus. In fact, the upper atmosphere of Uranus is very clear, so we are able to see down into a slightly lower layer where methane predominates. The overall chemical composition of the planet is also very different: Jupiter and Saturn are dominated by hydrogen and helium, which together make up around 90 per cent of their mass. Uranus is only about 20 per cent hydrogen and helium, with the rest being heavier elements such as oxygen, nitrogen, carbon and sulphur. Uranus and Neptune are, therefore, classified as ice giants,* to distinguish them from the two warmer gas giants.

Neptune, the Mystic

Neptune is another ice giant, although it has a rather bluer shade than the green-blue tint of Uranus. The colour again comes from methane, but this time in the upper atmosphere where it readily absorbs red light. We have not studied Neptune in much detail – with only one brief

. . . .

* In spite of the name, ice giants are still made of gas, but were given the name as they were originally made of icy constituents.

space probe flyby by Voyager 2 in 1989 – but the story of its discovery is an important one, as it was the first planet that was found by applying our knowledge of gravity and orbits in the solar system to predict its existence, rather than simply spotting it in the sky.

After the discovery of Uranus by Herschel, hunting for planets became a major pastime. For most, this simply involved scanning the sky night after night for unusual objects that moved relative to the stars, but others started looking for signs that a planet might be out there based on scientific calculations. In particular, it was noticed that the orbit of Uranus did not perfectly follow the predictions of Kepler's laws and Newton's gravitation. The Finnish-Swedish mathematician and physicist Anders Lexell (or Andrei Leksel, as he was known during a long period in Russia) suggested that this could be due to the gravitational influence of another, unseen planet. On 10 November 1845, the French mathematician Urbain Le Verrier presented the results of meticulous calculations that not only confirmed Lexell's proposal, but showed how the position of the new planet could be predicted. Less than a year later, on 31 August 1846, Le Verrier finished the laborious calculations and announced to the Académie Française the mass, orbit and current position of the postulated new planet.

Unfortunately, Le Verrier was rather an abrasive character and had few friends in France, and so nobody followed up his results. In frustration, on 18 September, he sent his predictions in a letter to Johann Galle of the Berlin Observatory. The letter arrived five days later, and the planet was found with the Berlin Fraunhofer refractor telescope that same evening, 23 September 1846, by Galle and Heinrich d'Arrest, within one degree of the predicted location.

With Neptune we reach the most distant known large planet, but the story may not end there. In recent years the same basic techniques used by Le Verrier and Adams have been applied to the orbits of other planets and dwarf planets (more on these below), and the evidence is growing of another ice giant beyond the orbit of Pluto. If true, it is only a matter of time before the predictions are precise enough to find this new planet and our solar system grows yet again. But even that may not be the end – there may be many more planets out there waiting to be discovered.

MINOR BODIES

The solar system doesn't just contain the large bodies of planets and moons; there are a lot of small objects left over. Dwarf planets like Pluto, asteroids and comets are all grouped under the general heading 'minor bodies'. As you will see, although they are not as dramatic as the giant planets, or as potentially life-bearing as some of the terrestrial planets and larger moons, they do have important stories to tell that will help us understand the origins and evolution of the solar system.

There are a range of minor bodies, and even more names for them, which can get quite confusing, so for clarity, here is a quick rundown of the main types:

- **Dwarf planets:** The largest minor bodies. They orbit the Sun like planets, but are smaller and sometimes have other similar objects in very close orbits (which full-size planets tend not to).

- **Asteroids:** Also known as 'minor planets', these are small, often irregularly shaped objects, mostly made of rock, in orbit around the Sun.

- **Comets:** Similar to asteroids but largely made of ice.

- **Meteoroids, meteors and meteorites:** Sometimes a small piece of space rock or dust hits the atmosphere of the Earth. When this happens, it is moving so fast that the sudden air resistance makes it burst into flames.

The different stages in this process have been given different names, which are often confused with each other (even by astronomers who should know better – I often have to look it up to check). The small piece of rock or dust is called a *meteoroid*. When it hits the atmosphere and bursts into flames, it creates a flash or bright streak that we call a *meteor* (also known as a 'shooting star', though of course stars are nothing to do with it). Small meteoroids will be burnt up in the air, but parts of larger ones may survive to reach the ground. This remnant is a *meteorite*.

Dwarf planets

The dwarf planets are probably the most controversial. Like many people, I learnt the list of nine planets as a child and was quite happy with Pluto as a small but important member of the elite. However, in 1992, the serendipitous discovery of another object in orbit beyond Pluto (called (15760) 1992 QB1)* led to a systematic search for more so-called trans-Neptunian objects or TNOs, which

. . . .

* It was originally suggested that the object should be called 'Smiley', but there was already an asteroid named after the astronomer Charles Smiley, so (15760) 1992 QB1 remains unnamed.

were expected to be found together in a big ring around the solar system called the Kuiper Belt.

Although it was named after him, the Dutch astronomer Gerard Kuiper did not actually discover or predict the Kuiper Belt, but he did predict something similar that might have existed in the distant past. His idea was that in the early history of the solar system, a giant messy ring of debris left over from the formation of the planets was orbiting just beyond the outermost large planets. Pluto formed out of this, but its gravity then scattered the rest of the 'Belt' much further out. However, this theory was based on the assumption that Pluto was similar in size to the Earth. When it was discovered that Pluto was in fact much smaller, and therefore had less gravitational effect on the remaining debris, it was realised that much of the debris could still be there, and Pluto, far from destroying the Belt, might just be one of many small 'planets' within it.

Soon many hundreds of TNOs were found, many of which were not much smaller than Pluto. Indeed, one – called Eris – could even be larger. Did this mean we had not nine, but many hundreds or even thousands of planets? Or were these a new class of object, of which Pluto was simply the first one to be found?

This led to an extensive and often acrimonious debate, but the conclusion was that leaving Pluto as a 'proper' planet was a hostage to fortune. Classification systems are intended to help examine a range of objects by grouping small numbers together, but with the potential for many thousands of new 'planets' to be discovered, the idea of a 'planet' could become confusing and useless. It was decided, therefore, to redesignate Pluto as a large member of a new class of dwarf planets – essentially like 'normal' planets but smaller. However, dwarf planets have some other important properties that make them quite distinctive. In particular, they are too small for their gravity to totally dominate the region around their orbit and so they can exist near to lots of other similar objects. In contrast, full-size planets will tend to either scatter or absorb nearby objects, or capture them and turn them into moons.

Of course, not everyone was happy and the 'Pluto killers' are still vilified by many. But in its new classification, Pluto has actually become more important, since instead of being just the smallest of the terrestrial planets, it has become the archetype and best studied of a whole new class. The flyby of Pluto by the New Horizons probe in January 2015 has revealed a vastly more complex and fascinating object than anyone – even the most devout Plutophiles – expected.

Much to the surprise of planetologists, who expected a cold, dead, largely featureless rock, Pluto shows signs that it is still evolving and changing, with 'ice volcanoes', mountain ranges and vast icy plains. It has also managed to keep a thin atmosphere and may even hide water just under the surface. The full implications of everything that New Horizons discovered will take many years to understand, but it is clear that the study of dwarf planets will grow enormously in the coming decades.

Asteroids

All the objects we have looked at so far are roughly spherical, since they were pulled into a ball by their own gravity during their formation. However, smaller objects have less gravity and so they have not formed into spheres: this is one of the things that marks the boundary between dwarf planets and asteroids. Asteroids, therefore, cover a very wide range of sizes, from rocks the size of a small country, down to small pebbles. There are also a lot of them: even if we only consider ones larger than a football pitch, there are still an estimated 150 million that are closer to us than Jupiter, of which less than 1 million have been found.

The vast majority of asteroids orbit in a rough band between Mars and Jupiter: the asteroid belt. They are probably the remains of a smallish planet that formed there, but was a bit too close to Jupiter and was torn apart by the constant tidal tugging. What they are made of supports this idea: the majority of asteroids are rocks similar to those found on Earth, but about one in twenty are almost pure iron or iron and nickel, and a similar fraction are a mixture of iron and rock. The rocky asteroids will, therefore, have come from the outer layers of the destroyed planet, the iron and nickel asteroids from deep in the core, and the mixed ones from the outer layer of the core.

The asteroid belt is regularly tugged and twisted by the gravity of Jupiter as it orbits past. This means that asteroids can quite easily change orbit, with some even leaving the asteroid belt altogether. Many are pulled towards Jupiter and end up in the same orbit. However, some are ejected from the Belt and fall inwards towards the Sun and, of course, us.

Asteroid impacts

About 65 million years ago, the dinosaurs still ruled the Earth. Probably the most successful class of large land animals to ever live, they had dominated the surface for at least 200 million years, but that was about to change, not due to a slow change in the climate or the gradual evolution of a stronger competitor, but almost certainly because of an asteroid impact.

The asteroid was about 10 km across and hit the ground at a place now called Chicxulub in Mexico, blasting a crater over 150 km wide and about 19 km deep. Obviously anything nearby would have been obliterated by the impact itself, but the real damage was caused by the after-effects: vast quantities of dust were thrown into the air, and the shockwave and radiated heat started global forest fires, adding smoke and ash to the dust. Any land animals that survived the impact, the fire and the loss of food and habitat would then have had to survive many years of intense cold as the dust blocked out the light and heat of the Sun. Our distant mammalian ancestors were able to make it through, but around three quarters of all animal and plant species were made extinct, including the last of the dinosaurs.

Of course, from a purely selfish point of view, this was a good thing, since it left the way clear for us to evolve into our dominant position, but there is no reason to suppose that this is the last time the Earth will be hit by a large asteroid. Globally catastrophic events are very rare – they might happen only once every 100 million years or so – but smaller asteroids do impact more frequently and can still cause major amounts of damage. They may not threaten our extinction, but they might put our civilisation at risk. In 1908, for example, an explosion over Tunguska in Siberia completely flattened a forest the size of London – although Tunguska is so remote that it was some time before anyone noticed. But we would expect such events roughly every century, and there is no telling where it might happen next time. Impacts with the same explosive energy as the Hiroshima atomic bomb occur every year or two, although most detonate over oceans or deserts and are only detected by sensitive instruments.

For most of human history, this has not been something that was worth worrying about, but things have now changed in two important ways. Firstly, the growing human population means that an increasingly large area of the Earth's surface is inhabited, and so the risk of a major disaster is increasing. Secondly, and most importantly, now we might actually be able to do something about it, if we know it's coming.

Since the early 1990s, an international consortium of scientists, engineers and enthusiastic amateurs have been scanning the skies for asteroids, plotting their paths and calculating orbits that allow us to predict possible hazards. Although they're still in their early days, these Spaceguard programmes around the world have already identified and logged tens of thousands of so-called near-Earth objects (NEOs). Of course, there is no point identifying hazardous asteroids if we can't do anything about them, so alongside these discoveries, techniques for protecting us from asteroids are being devised.

You might think that the obvious thing to do if an asteroid is heading towards us is to use nuclear bombs to blow it up, Bruce Willis style. Unfortunately, that will almost always be a bad idea. While you might break up the asteroid, it would be moving far too fast to significantly change the direction of the pieces, and so they would all still hit us. It is rather like the difference between being shot with a rifle and being shot with a shotgun – far better not to be shot at all.

Most techniques try to deflect the asteroid, nudging it just far enough off course that it misses the Earth completely. The further away the asteroid is, the smaller the nudge needed, so these techniques rely on plenty of warning, which is what the Spaceguard programmes should give us.

The exact technique chosen will depend quite a lot on the nature of the asteroid. Some asteroids are very tough and solid, and so simply ramming them with a rocket might do the job, but if the asteroid is even slightly fragile, that could just lead to the shotgun problem again.

A better approach is to gently push against the asteroid for an extended period. Since asteroids usually spin and are knobbly and irregular, this is very hard to do with a rocket (and also requires vast amounts of fuel), but there are other ways of 'pushing'. One of the simplest ideas is the 'gravity tractor'. This works by using the gravity that exists between all things. We think of gravity as being the province of huge objects like planets and stars, but everything has its own gravitational pull – it is just very small for very small objects. So there is a tiny pull between you and this book, it is just too small for you to feel it. Out in space, however, there is no air resistance or anything else to counteract the force of gravity, so even very small gravitational pulls will gradually have an effect. The gravity tractor is essentially just a rocket that is put into an orbit right next to the asteroid. Without even touching it, the gravity of the rocket would slowly pull the asteroid towards it (and vice versa), making a tiny change to its orbit that, given long enough, should make it miss the Earth.

Of course, sometimes there might not be enough time for the tractor to do its job, so a rather more dramatic approach would be needed. One idea does involve a large nuclear bomb, but not to blow the asteroid up. If the bomb is detonated not on the asteroid, but just next to it, there would be no direct impact on the asteroid as there is no blast or shockwave in space. However, the heat from the explosion can travel without air, and so one side of the asteroid would become very hot. Some of the surface would be vaporised, and the gas created would 'squirt' into space, acting like a crude jet. Since this would only happen on the side facing the explosion, the jet would push the asteroid sideways, changing the orbit.

Another technique — and my personal favourite — involves nothing more than a lot of white or silver paint. Another tiny effect that is all around us, but which is too small for us to feel, is 'photon pressure' — the gentle push of light hitting things. Asteroids are being very gently pushed by sunlight all the time, but as they spin and tumble through space, the effect is largely cancelled out. However, if we could paint one side of an asteroid in a much lighter colour, more light would reflect off that side. This would not directly push the asteroid, but the *spinning* of the asteroid would change. Nothing can change in total isolation, however, and one of the fundamental laws of physics is that the total amount of circular movement in a system – whether it is 'spin' like a spinning top or rotation in orbit around something – stays fixed.* So, if the spin of the asteroid is increased, some other kind of circular movement must be decreased to compensate. The only other circular movement is the orbital motion, and so by changing the tumbling spin, you also change the orbit.

There are many other such techniques being considered and developed, and although none have been tried in earnest, perhaps by planning ahead and searching the skies we are giving ourselves a way of avoiding the same fate as the dinosaurs.

Comets

Anyone who has ever seen a comet spread its tail will never forget it. I was lucky enough to get some beautiful views of Comet Hale–Bopp in 1997, and was captivated as it appeared to arch across the evening sky. It therefore comes as a surprise to many that comets are actually quite small and little more than balls of dirty ice. For most of the time, comets are very far from the Sun. The majority are found in the Oort Cloud, which is a loose shell at least

----

* This is called the conservation of angular momentum.

10,000 million km across, surrounding the solar system and containing billions of comets.

Occasionally – perhaps because of a close encounter with another comet, perhaps because of a gravitational nudge from a passing star – a comet falls into an extremely elongated orbit, skims through the inner parts of the solar system, around the Sun and back out towards the Cloud. This long orbit may take 100,000 years, and for almost all of that time the comet is small, dark and essentially invisible. However, when it gets closer to the Sun, the surface of the comet heats up and some of the ice (made from both water and carbon dioxide) begins to evaporate, surrounding the small comet with a much larger gas cloud called the coma, which is then swept up by the solar wind and the pressure from sunlight and streams out as a tail, finally becoming visible to us. Eventually, of course, all the ice will evaporate away and the comet will be no more. For most comets that will take many orbits, but some, like Icarus, fly too close to the Sun and are destroyed.

The most well-known comet is a bit different. Halley's Comet does not take many thousands of years to orbit, but famously returns every 75 years. This regular behaviour was first noticed by Edmund Halley, in 1705, who realised that comets seen in 1682 and 1531 had very similar orbital characteristics and were likely to be the same comet appearing twice. He used his calculations to predict the return of the comet in 1758, which it did, thus immortalising Halley. Such 'short-period' comets probably started out as long-period visitors from the Oort Cloud, but were deflected and captured by the gravity of a close approach to a planet (usually Jupiter or Saturn).

Another comet that had a moment of fame, in 2014, was 67P/Churyumov–Gerasimenko (usually abbreviated to just '67P'). With an orbital period of just over six years, it returns even more frequently than Halley, which made it ideal for an ambitious plan to not just study it from afar, but actually send a spacecraft to it and even land a small robot there. After its launch in 2004, the Rosetta satellite spent ten years flying to 67P and then matching its orbit. The images and data flowing back helped to revolutionise our understanding of comets, but it was the small lander – Philae – that was the most ambitious part of the project. On 12 November 2014, astronomers all over the world were glued to their computers as Philae detached from Rosetta and flew towards the comet. I was with a number of colleagues watching its progress on a big screen in our staff room as Philae got closer and closer, finally approaching the planned landing site at the precise required speed of just 1 metre each second (a slow walking pace). And then – landing! Watching the celebrations of the Philae team on the live broadcast from mission control was a joy, and brought home the incredible work behind a project like this. From start to finish, a mission might take several decades of work – the focus of an entire career – all of which converges on that brief moment of success or failure.

Unfortunately, in this case the celebrations were short-lived. After the initial touchdown, communication was lost and the expected data did not arrive. It was only two years later, when Philae was finally spotted by the cameras on Rosetta, that the whole story became known. The landing had indeed gone to plan, but two 'harpoons' designed to grapple Philae to the surface of the comet failed to fire and the tiny lander bounced a couple of times. Even then all would not have been lost, as it was essentially undamaged, but it happened to end up in the shadow of a large cliff – out of sight and unable to charge its solar panels, it couldn't communicate except in occasional brief bursts.

In spite of this, the Rosetta mission as a whole was hugely successful, ending in September 2016 when the main probe was deliberately crashed into the comet after supplying vast quantities of data on the composition and origin of the comet that will take many years to unravel. Meanwhile, 67P itself continues on its brief orbit out around Jupiter and back towards the Sun – perhaps to be the target for another mission in the future.

One additional surprising gift we get from comets is beautiful meteor showers. Meteors, or 'shooting stars', are obviously not really stars, but small pieces of space dust hitting the upper atmosphere at high speed and burning up because of the friction. Meteors can be seen a few times an hour on most clear, dark nights, but sometimes they become much more frequent, and for a few nights you might see as many as several a minute. This happens because the Earth is passing through a region of space that a comet has previously been through, leaving a trail of dust from its tail and providing many more particles to make meteors. There are several such showers each year, but perhaps the best to look out for are the Perseids in August, the Draconids in October and the Leonids in November.

From the Sun and the giant planets, right down to the smallest dust particle burning up as a meteor, everything has its fascinating story to tell. Putting all these stories together can help us form a much greater understanding of the solar system as a whole – its origins, how it has evolved over the years and perhaps what the future holds.

For example, by observing all the different objects – their composition, behaviour and so on – we were able to provide a theory for a burning question: how was the solar system made?

FORMATION OF THE SOLAR SYSTEM

Everything in the solar system started out as a single, giant cloud of dust and gas floating between the stars. Left alone, this would not have changed much, but billions of years ago, something happened – perhaps a nearby star exploded, or another star passed very close by – and the cloud started to collapse.

As the cloud collapsed, it started to spin faster and faster (as happens to anything that is spinning and changing shape or size; spinning ice skaters, for example, speed up when they pull their arms in). As the cloud started to spin faster, it formed into a large, spinning disc. The centre of the disc became very hot and the pressure increased as gravity pulled in more and more of the gas. Eventually, a large amount of the gas was concentrated in a very hot, dense ball, which became so hot and compressed that nuclear reactions started in the centre and it turned into a star: the Sun was born.

Further out in the disc, where it was relatively cool but still becoming fairly dense, the gases started to condense and form more complicated molecules of gas and small particles of dust. Close to the new Sun it was too hot for the lighter gases to survive, so only the heavier elements condensed. The particles of dust were pulled together by gravity and started to form larger and larger lumps, which had even more gravity and so pulled in other lumps, slowly forming the rocky planets. These continued to grow until they had used up all the dust nearby. Further out from the Sun the more volatile gases survived and dominated, and were pulled together by gravity to form the much larger but less dense giant planets. And even further out, where the gases were rarer, rocky objects again dominated.

At this point, the solar system was a very busy place, with a lot of new objects flying around and interacting with each other. This meant a lot of collisions and cratering, and the lighter objects tended to be thrown about by the gravity of heavy ones, so the giant planets probably kicked quite a lot of smaller objects out beyond the orbit of Neptune.

Finally, the Sun went through a brief but very energetic period in its development called the T Tauri phase. At this point it produced a strong solar wind which swept away any remaining gas and dust, leaving the relatively stable system that we see around us now, which has remained largely unchanged for several billion years.

This explanation accounts for most of what we can observe in our solar system. With the rocky planets close to the Sun, the gas giants further out and the assorted remnants and debris making up the dwarf planets, comets and asteroids, it all fits together. However, there is one serious problem: the model was created to explain the solar system, so it is not really a surprise that it does so. It accounts for everything we have observed, but without being able to test the theory on another example, a theory it remained. Until we started to discover other planets orbiting other stars.

EXOPLANETS

For a long time, there was no proof that planets existed outside of our solar system, but since they were first discovered, the search for these extra-solar planets (or exoplanets) is probably the fastest growing area of astronomy. In less than a handful of decades, we have gone from wondering if the solar system might be unique, to having catalogues of many hundreds of other planetary systems.

This rapid change has resulted from an improvement in the technology and techniques needed to find them.

Finding exoplanets is very, very hard. Planets do not shine by their own light like stars do, but only reflect light from the star they orbit. This gives us two problems: distant planets are going to be faint, as they are both small and dim; and we can only hope to see them at all if they are very close to a bright star, which will nearly always mean they are lost in the glare from their star. It is rather like trying to spot a single match held in front of a massive spotlight. This means that finding most exoplanets by getting pictures of them is virtually impossible and other sophisticated methods have to be used. These techniques do not look for the planet directly, but try to find the effect that the planet has on its star – perhaps by making the star wobble slightly, or by occasionally blocking some of the star's light as the planet orbits in front of it.

This is why exoplanet hunting had a slow start – there was a lot of resistance to allocating the time, effort, money and telescopes needed to do it. There were good reasons to be sceptical: first of all, larger, more massive planets are much easier to detect, so finding Jupiter-like planets might be feasible, but not anything as small as the Earth. Secondly, to be confident that you have found an exoplanet, you need to observe the system for at least two full orbits. Since we knew that large, gas giant planets could not form close to stars, where it is very hot, we would have to study stars for decades before anything interesting could be found. Given the rapid advance of technology, it made sense to wait for the development of better telescopes and cameras that could find smaller, faster-moving planets.

The discovery in 1992 of the very first planet outside the solar system might have been enough to convince the sceptics, but it was something of an unusual case. The exoplanet was found by Aleksander Wolszczan and Dale Frail in orbit around a pulsar − a rare, rapidly spinning, very dense, dead star that can only be seen by radio telescopes. Although an important discovery, it was sufficiently odd that it did not really say anything useful about the possibility of planets around 'normal' stars, and so the problems remained.

However, some teams were not convinced and went ahead with the hunt anyway, building high-precision instruments and observing nearby stars night after night. One such project was set up by Michel Mayor and his student Didier Queloz at the University of Geneva. They had planned to observe for many years and slowly build up the data they needed to find slow-orbiting but massive gas giants. Early on in the project, however, Queloz was working with some observations to practise his data analysis skills and was astonished to find evidence for a planet. As expected, the planet was large – probably about the same size as Jupiter – but its orbit was not the years or decades he had expected, but just over four days! Not only had Queloz and Mayor found their first exoplanet, they had turned our understanding of planets on its head. Any planet that large had to be a gas giant – a rocky planet would collapse under its own gravity – but at that distance from its star, there was no way that the light gases needed to make a giant could possibly have formed.

It was not long before other similar exoplanets were found, and soon there were catalogues of these so-called 'hot Jupiters', which were too common to be explained simply as anomalies. Our theory of planetary formation based on our solar system could not account for them, so it had to be wrong (or at the very least incomplete).

There are various theories being suggested to account for these planets – for example, the idea of 'migration', in which a planet's orbit gradually drifts inwards or outwards from its star. In this case, the planet could form further from the star and then at a later point move closer. But, although we're constantly learning more, this is still not something we fully understand, and is also not compatible with our own solar system, where there is no obvious evidence of the migration process.

Nevertheless, this is an ignorance we should cherish. We may seem to have gone backwards, from a nice clear, consistent model that explained what we saw around us, to one where everything is more complicated and leaves several important questions unanswered. But this is one of the exciting if sometimes frustrating aspects of astronomy – theories and assumptions can always be challenged as new evidence and observations come to light, leading the way for a whole new approach. The study of exoplanets is having many benefits for our understanding of the universe, but it is perhaps what it is telling us about our own solar system that is most important.

4

STARS

Planets are obviously important to us — we live on one, after all — but it is easy to ignore how much stars dominate the universe. In our modern world of street lights and pollution, the astonishing sweep of the night sky and the sheer *number* of stars is easy to forget. But away from the lights and smog, with hundreds of stars twinkling in the sky and the Milky Way arching overhead, it is impossible not to be lost in wonder. From the earliest recorded history, humanity has tried to understand and explain the stars. Many cultures developed rich mythologies that peopled the stars and constellations with gods, heroes, devils or just the animals and plants around them. For many, the night sky and heaven were one and the same – the beauty and mystery of the stars perhaps overcoming the fear of the dark.

As an astronomer, my view of the stars is in some ways very practical: stars dominate the sky, and so figuring out what they are and how they work is obviously important. However, as we learn more about them the wonder only increases, and when I'm lucky enough to see a truly dark, clear sky, I am as captivated as anyone.

CHARACTERISTICS OF STARS

Broadly, all stars look pretty similar – they are all tiny dots that twinkle and flicker in the night sky. But there are some key differences that can tell us a lot about them, their history and their evolution. Most noticeably, some are brighter than others; most stars can only be seen on the darkest, clearest nights, but a few, such as Sirius or Vega, are bright enough to be seen not long after sunset, when twilight still colours the sky. The distance of a star from the Earth affects how bright a star appears to us, but its brightness is also affected by how luminous it is – a result of both its size and temperature. This is one of the things that makes distance measurement one of the trickiest problems in astronomy: does a star look bright because it is close, or because it is luminous? The extent of the problem is clear from the closest known star to the solar system. Called Proxima Centauri, it is a mere 4.2 light years (40,000,000,000,000 km) from the Earth, but it was not known about until 1915, when it was discovered by Scottish astronomer Robert Innes working in South Africa. That the closest star could remain undiscovered for so long is a testament to its faintness – it appears more than 100,000 times fainter than Sirius, the brightest star in the night sky, which is twice as far away.

Luckily, we can learn about the variations in the brightness of stars thanks to those that appear in clusters. On a clear winter night in the northern hemisphere, the constellation of Orion is probably the most easily spotted feature, with its 'belt' of three stars being particularly recognisable. If you allow your eye to follow the line of the belt up and to the right you should spot an unusual object: not a star, but an extended, rather fuzzy blob. This is the Pleiades or Seven Sisters, and with even a very small pair of binoculars the 'fuzz' separates out into a tightly knit group of stars (though whether you can see six, seven, eight or even more 'sisters' does depend on your eyesight, your binoculars and the weather). Although there are many stars spread across the sky and at all distances, it is virtually impossible to have so many similar stars appear so close together just by chance, so they must be a real group of stars: a cluster. Since stars that cluster together must all be at the same distance from us, any visible difference in brightness must be intrinsic to the stars themselves.

Another interesting and useful aspect of a star we can look at is its colour. The colours of stars are hard to see with the naked human eye, which doesn't differentiate colour well in the dark. However, with the invention of the telescope, we have been able to gather enough light from a star for the wide range of colours to become clear. A good photograph of the night sky will show stars of many colours: red, orange, yellow, blue, violet. Even in a single constellation, the variation in colour can be dramatic: Betelgeuse in the top left of Orion is orange-red, contrasting strongly with the fierce blue of Rigel at the bottom right.*

These colours can be easily explained by differences in temperature. The mechanism that produces the light we see is (rather confusingly) called black-body radiation. The

* See plate section.

idea behind this is very simple – hot things glow, and the colour they glow changes as they get hotter. It was Max Planck in the twentieth century who explained how hot things produce light, and with it the relationship between temperature and colour – in other words, exactly how the range and mix of different colours of light shifts towards blue as the temperature increases. Cooler temperatures – say up to a couple of thousand degrees Celsius – are at the red end of the spectrum; indeed much of the 'light' produced is not visible at all, but infrared (which is why electric fires feel hot). Higher temperatures move up to the blue end, and even past it into the ultraviolet.

So, the element of an electric fire glows a dull red, while liquid metal being poured into a cast is yellow or even 'white hot'. And in the same way, the Sun has a surface temperature of around 6,000°C and glows yellow. Cooler stars might have surfaces of just 3,000°C – similar to the element of an electric fire, and so the same red colour – and the very hottest stars have surfaces in excess of 40,000°C, making them blue-white or pale purple. Being able to observe the colour means we can estimate the surface temperature of a star.

Interestingly, though, no stars look green. Intermediate temperatures of a few thousand degrees Celsius are near the middle of the visible spectrum where the green light is. However, the spectrum itself is very broad so there is still a lot of red, yellow, blue and purple light included and our eye does not see just the green, but a yellowy-white colour. Stars similar to the Sun have surface temperatures that are green on the Planck spectrum, but they look yellow — so, no stars are green.

Knowing both the surface temperature and brightness of the stars can tell us a lot about them. When the two are plotted on a graph (first done around 1910 by two astronomers working independently: Ejnar Hertzsprung and Henry Norris Russell, resulting in the Hertzsprung-Russell diagram, or HR diagram), a clear pattern emerges between temperature and brightness. The majority of very luminous stars are hot and blue, medium-luminosity stars have intermediate temperatures (like the Sun) and the faintest stars are cool and red. And so almost all stars sit on a narrow line that runs diagonally across the graph — this is called the 'main sequence'.

The main sequence covers a wide range of brightnesses and temperatures. Where a star appears on the graph is down to its mass — the brightest stars are tens of times more massive than the faintest. The more mass a star has, the greater the force of its gravity and the more heat is

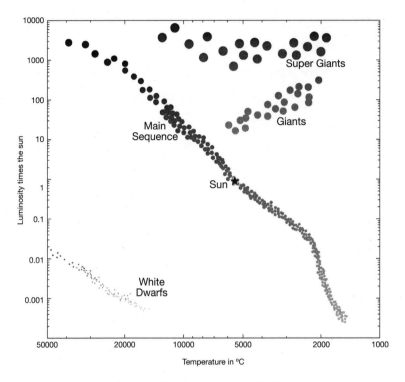

A schematic Hertzsprung-Russell diagram.

needed to balance it. So, the greater the mass, the brighter and hotter the star.

However, there are a scattering of exceptions, with some stars being both very luminous (much brighter than the Sun) but also quite cool and red, for example. This is because, as we've seen, the brightness of a star is related to its size as well as its temperature, so stars that are cooler but bigger can also be very bright, while some stars that are very hot but small can be faint.

We have ended up with three distinct types of star: main sequence stars, which go from small, cool, faint, red stars, right up to very large, hot, bright, blue stars; a smaller number of very bright, large but cool and red stars, very sensibly known as red giants; and small, faint but very hot stars called white dwarfs,* which we detected as telescopes got bigger.

Identifying these three broad types of stars has given us a number of clues about the nature and evolution of stars, and led us to a more complete understanding of their life cycle: how they form, change, evolve and, eventually, 'die'.

THE STORY OF A STAR

We've already seen an outline of how stars form in Chapter 3. Vast clouds of gas and dust are quite common in space, and although they are quite stable, thanks to their own heat holding them up against the squeezing force of

. . . .

* Although the plural of 'dwarf' is usually given as 'dwarves', this is a relatively recent usage popularised by J. R. R. Tolkien, and astronomers have stuck to the older usage, although nobody seems to know why.

gravity, that stability can be disturbed by the shockwave of a nearby exploding star, or perhaps by a close encounter with another cloud. When that happens, the cloud will break into smaller (but still very large) clouds which can no longer hold themselves up against gravity and will collapse. As a cloud collapses, it heats up, particularly in the very centre, which will also become very dense as gravity continues to squeeze and crush. By astronomical standards, this is a very fast process, taking just a few tens of thousands of years, but there comes a point when the rapid collapse is stopped by a new force: heat from the nuclear fusion that kicks off in the hot, dense core, with hydrogen fusing to create helium and generating vast amounts of heat as it does so. A star is born.

At this point the new 'protostar' is still surrounded by the remains of the cloud of gas and dust that it formed from, so we cannot see it clearly – all we can see is the glow from the cloud itself, which is heated up by the new star and glows brightly in the infrared, but produces very little visible light and so is not found anywhere on the HR diagram. However, as the protostar settles down, it goes through a brief but violent period called the T Tauri phase, where it produces a very strong 'wind' of particles which blow away the last remnants of the cloud, slowly revealing the star in the centre. From our point of view, this means that the light from the star becomes increasingly visible, so it appears to get gradually hotter and brighter, moving into the cool-and-faint corner of the HR diagram and then shifting up and across until we see the star in all its glory as it joins the main sequence.

The stars on the main sequence do not move up and down the line, but stay put for long periods of time. As long as there is plenty of hydrogen in the centre of a star, where the fusion occurs, everything will remain in balance and the star will be almost entirely unchanging. However, eventually the hydrogen near the hot, dense core will run out and the nuclear reactions will stop. For a star like the Sun, this takes about 9 billion years (making the Sun middle-aged at about 4.5 billion years old) and, quite suddenly, the balance between gravity and heat is broken, and gravity is the winner. Then things start to change.

Red giants

The core of the star shrinks as gravity squeezes it, but this brings fresh hydrogen nearer to the centre, where it can start to fuse into helium once more. This creates a 'shell' of nuclear reactions and heat which has a strange effect on the outer parts of the star. Although the core is shrinking, the new shell of energy heats up the inner layers of the star, making them expand. These in turn push the outermost layers up, but they are not being directly heated and so they actually cool down. The net effect is that the star appears to get much bigger, but also cool down: it moves off the main sequence on the HR diagram and becomes a red giant.

When the Sun reaches this stage, it will not be a good time for the Earth. Although it is hard to predict exactly what will happen, we do know that the Sun will get so large that the Earth will either skim its surface or be swallowed up inside. Mars, however, should become relatively pleasant, so our very distant descendants will at least have the option of moving there (assuming they've managed to overcome the obstacles mentioned in Chapter 3).

While the outer parts of the star are busy growing (and perhaps swallowing up nearby planets), in the core, far below the outer layers that we can see, things become much more complicated. The helium is squeezed by gravity, pushing the constituent parts of the atoms much closer together than is normally possible.* When

* Atoms are made up of smaller particles called protons, neutrons and electrons, held together by several different forces.

they get very close, they start to push back. Normally, gases expand when they get hot, but this compressed 'gas' does not, so the temperature can get very high without any change in the size or density of the core. Eventually it will get hot enough that a new kind of nuclear fusion can occur, this time with helium as the fuel, but because of the strange, unchanging pressure, the core does not expand and cool, but just gets hotter, so the reactions get faster, and so even hotter, and so on to a runaway nuclear reaction that drives the core to incredibly high temperatures. Eventually, however, the temperature gets so high that the other parts of the atoms crushed into the core can produce a pressure of their own, making the core expand and turn back into something much more like a normal star.

This runaway reaction is called the helium flash, and it rapidly changes the internal workings of the star, but very little − if any − of this is seen at the surface. This is because, although the helium flash is very energetic, it only lasts a short length of time − a matter of minutes − and the change does not reach the surface layers. However, at the end of the flash, the nuclear fusion of helium is still continuing in the core, together with the shell of fusing hydrogen around it, and, for a while at least,

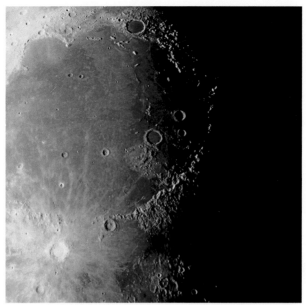

A. Newsam/Liverpool Telescope/LJMU

Part of the Lunar terminator with mountain tops illuminated by the setting Sun.

A.Newsam/Based on data collected at Subaru Telescope, operated by the National Astronomical Observatory of Japan

A small part of a deep observation from the Subaru Telescope, operated by the National Astronomical Observatory of Japan. Apart from a few stars (which look round and relatively sharp-edged) almost everything in this image is a galaxy, most of which had never been seen before.

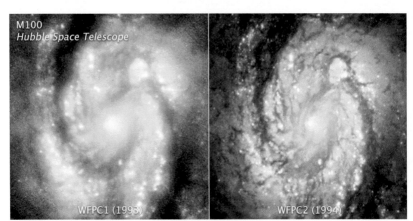

M100
Hubble Space Telescope

WFPC1 (1993)

WFPC2 (1994)

NASA, ESA, STScI and Judy Schmidt

The difference in Hubble Space Telescope images before and after the 'fix' (see page 25).

NASA, ESA, A. Simon (Goddard Space Flight Center), and M.H. Wong (University of California, Berkeley)

HST observation of Jupiter and its storms (see pages 26 and 61).

The 'Pillars of Creation'. An HST observation showing giant clouds of dust and gas sculpted by the heat from nearby, new stars.

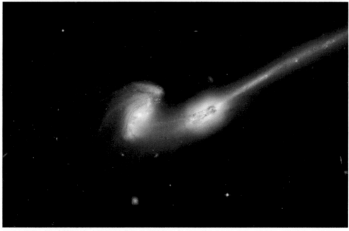

The 'Mice' pair of interacting galaxies, also known as NGC 4676, passing so close to each other that the tidal effect of their mutual gravity is tearing them apart.

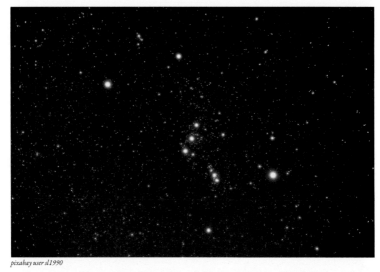

pixabay user sl1990

Orion, with red Betelgeuse clear at Orion's 'shoulder' (see page 102).

NASA, ESA and Allison Loll/Jeff Hester (Arizona State University). Acknowledgement: Davide De Martin (ESA/Hubble)

The Crab Nebula – the remnant of a vast supernova explosion.

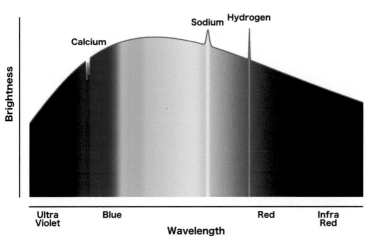

A simplified spectrum of a star (see page 36). At the top is the spectrum as it might look to the eye. Below is the data as a graph. The overall shape is caused by the 'glow' of a hot star – in this case about the same temperature as the sun. You can see how the 'light' continues into the ultra-violet and infra-red, which the camera can detect, even if our eyes cannot. To the right, in the red area, is a strong emission line made by hot hydrogen. Next to this is a slightly weaker but wider orange line. This is caused by sodium and is not from the star, but the faint glow from distant street lights. Finally, to the left there are a pair of darker absorption lines created by calcium.

Stuart Kettle

The Milky Way over La Palma in the Canary Islands.

ESA/Hubble & NASA. Acknowledgement: Stefano Campani

A globular cluster of stars (NGC 121).

ESA/Hubble & NASA, P. Cote

Elliptical galaxy Messier 59 observed by the HST.

Barred spiral galaxy UGC 12158 observed by HST.

European Space Agency & NASA, Acknowledgements: Project Investigators for the original Hubble data: K.D. Kuntz (GSFC), F. Bresolin (University of Hawaii), J. Trauger (JPL), J. Mould (NOAO), and Y.-H. Chu (University of Illinois, Urbana), Image processing: Davide De Martin (ESA/Hubble), CFHT image: Canada-France-Hawaii Telescope/J.-C. Cuillandre/Coelum, NOAO image: George Jacoby, Bruce Bohannan, Mark Hanna/NOAO/AURA/NSF

The Pinwheel, a spiral galaxy.

The Cosmic Microwave Background (see page 179) over the whole sky as observed by the WMAP satellite. The darker, bluer areas are cooler, and the yellow and red hotter, although the largest difference in temperature is less than 1 part in 1000.

The Hubble Ultra Deep Field (see page 182). This is one of the deepest observations ever taken, and shows galaxies so far away the light from them has taken billions of years to get here.

the star appears to be normal again. The outer, visible parts shrink and heat up and the star moves back to the main sequence.

This doesn't last though. Helium is a less efficient fuel for fusion than hydrogen, and there is less of it in the star to start with, so the helium-burning does not last as long: for the Sun, this phase will last about 100 million years, far less than the 9,000 million years of its hydrogen-burning phase. When the usable helium is exhausted, the core will collapse again and the star will go into a new red giant phase.

White dwarfs

At this point, for an average-size star like the Sun, the temperature in the centre will never get hot enough to start nuclear reactions with any other kinds of atom, and so there will be no more sources of heat: the star is, in effect, 'dead'.

The outer parts of the red giant will continue to slowly expand, cooling as they do so and gradually becoming more transparent. Eventually they will cease to look like a star at all, but will become a large, roughly circular cloud of various gases called a planetary nebula (these are so called not because they have anything to do with a planet

– they are usually much larger than our entire solar system – but because they looked a bit like planets through early telescopes). Eventually the nebula itself will fade away, leaving the remains of the core of the star sitting alone.

This remnant is a very strange object. It is not normal matter at all, as the atoms have been squeezed together so much by gravity that they have formed a very dense material called 'electron degenerate matter': a piece of this the size of a grape would weigh as much as an elephant. As well as being very dense and small, this remnant is also very hot (maybe 100,000°C) and so the star looks very blue, but also very faint – it is now a white dwarf.

Nothing much more will happen to a white dwarf. It has no new sources of heat or energy to make it grow, but gravity cannot beat the strange degeneracy pressure and so it also will not shrink. It will just very slowly cool down until eventually it stops emitting light and heat altogether, becoming a dark 'black dwarf'. This cooling down will take many billions of years – far longer than the universe has even existed so far – so there are no black dwarfs for us to try to find, but the eventual cooling is driven by the fundamental laws of physics and is as inevitable as death and taxes.

Neutron stars

That is not the end of the story for all stars, however. Stars that are more massive than the Sun will come to a more violent end after the red giant phase.

At first sight, it might seem that the more massive a star, the longer it will last on the main sequence; they have much more hydrogen as fuel and so their nuclear reactions should continue for longer. However, this does not take into account the much higher luminosity of these stars. While a star towards the hotter end of the main sequence will have around 20 times the hydrogen of the Sun, it will be more than 30,000 times as bright, so it will use up its fuel much more quickly, lasting just a few million years. When the fuel in the core runs out, the star will at first follow a story similar to the one above, up to the red giant phase.

The extra mass in these stars means that their gravity is stronger and their cores can get a lot hotter, so increasingly heavy atoms can be used as nuclear fuel. This leads to an 'onion skin' structure inside the star, with an outer cloud of hydrogen, and then successive layers of different elements burning in 'shells': hydrogen is the furthest out and then moving inwards there is helium, carbon, oxygen, neon, magnesium and silicon. However, at this point the process stops, as the next element would be iron, which is too heavy to fuse together without an additional energy source. Once all the energy sources have been used up, there is no new source of pressure to hold up the core against gravity, and so it collapses very rapidly; it can shrink down from the size of the Earth to the size of a small town in a matter of seconds. The other layers suddenly have no support from below and they too fall inwards, hitting the tiny core and 'bouncing' back. This bounce produces a vast explosion called a supernova.

The energy of a supernova is mind-boggling. In astronomy, we are used to dealing with large numbers, but the energy of a supernova is something special. In just a few seconds, more energy is released than the Sun will produce in its entire 9 billion years on the main sequence, making the star so bright that some supernovae have even been seen during the day on Earth. Unfortunately these 'daytime' supernovae are quite rare – the last was about 500 years ago. This is odd, as we expect there to be a supernova in our own galaxy roughly every fifty years and most of them should be very bright. However, nearly all of them are hidden behind or inside giant clouds of dust and gas and so we do not see them at all – in fact, there has not been a single observable supernova in our own galaxy since the invention of the telescope, which is very frustrating.

As well as making the supernova look impressive, there are other important consequences of this vast release of energy. The elements heavier than iron now have the energy they need to fuse, and that is exactly what happens in the shockwave of the supernova explosion, where a wide range of new, heavy elements are created. In fact, this is one of the few places in the universe that these heavy elements can be made. Of the ninety-odd naturally occurring elements, two (hydrogen and helium) were made in the early universe, three (lithium, beryllium and boron) can be made by collisions between particles in space, and all of the rest are made by stars – either inside them or during supernova explosions. This means that almost all the different atoms in your body were made by stars and spread through space by their 'deaths'; we really are made of stardust.

While the outer parts of the star are being blown apart, transformed and dispersed by the supernova explosion, the core is also being battered. The force of the bouncing outer layers combines with gravity to further crush it. At this point, if the remains of the core are around two or three times the mass of the Sun, the collapse is stopped when the atoms are crushed into their smallest possible state. This means that the core of the star has become a ball of neutrons about 20 km across: a neutron star, the densest solid object in the universe. Where a piece of white dwarf the size of a grape would weigh as much as an elephant, a similar piece of a neutron star would weigh more than all the people on Earth combined!

Being so small, neutron stars are far too faint for us to see directly, but we have found some due to their very strong magnetic fields. These magnetic fields produce beams of radio waves, which sweep around as the neutron star spins, rather like the light on a lighthouse. If we are lucky enough for the beams to sweep past us, we see a flash of radio waves which repeats with each spin. This makes these particular neutron stars visible and they are called pulsars.

Pulsars

Pulsars were discovered entirely by accident. In the late 1960s, Cambridge astronomer Antony Hewish designed and – with the help of some students – built a very unusual new radio telescope in a field on the outskirts of Cambridge. It was not the giant 'dish' that you might expect, but a mass of over 1,000 poles sticking out of the ground, with around 190 km of wires strung between them (and a small flock of sheep to keep the grass short).

The telescope was designed to look for rapid changes in brightness called 'scintillation', which are caused by radio waves from distant objects passing through clumps and patches in the solar wind.

The data rolled off the telescope as very long pen-charts – basically a squiggly line up to 30 metres long each day – which had to be studied by eye daily. This task was the job of Hewish's PhD student Jocelyn Bell (now Bell Burnell), and in July 1967, Bell noticed something very unusual in the charts, unlike anything she had seen before. What made the discovery so special was that the signal was both very fast (repeating almost every second) and very regular and precise. All astronomical objects known at the time could only vary slightly randomly, so something regular seemed to be a sign of intelligence. And if it wasn't from the Earth, could it be aliens? Like any good scientist, Bell was initially sceptical and assumed that it was interference from a local ham radio or something similar. However, after many weeks' work, she was able to show that the source of the odd signal moved with the stars across the sky – it was not fixed on the Earth like a radio transmitter.

Still sceptical, Bell and Hewish gave the signal the tongue-in-cheek name 'LGM-1' (Little Green Men 1), but as time went on and no sensible astronomical explanations were found, the team started to seriously consider whether this might be evidence of another civilisation. As Bell herself said later, 'If one thinks one may have detected life elsewhere in the universe, how does one announce the results responsibly? Who does one tell first?'

Carrying out further work over the following months, Bell – with help from Hewish – was able to find three or four other similar 'pulsing' sources in different parts of the sky, which made an astronomical explanation much more likely (rather than a galaxy-spanning civilisation that only had a few outposts). However, it was still many years before spinning neutron stars with their sweeping beams of radio waves were shown to be the cause.

Of course, although aliens had been ruled out by the time the results were published, the press was far more excited about that than about the reality. And when they discovered that a woman was involved in the discovery (even rarer then than it is now), the press interest grew considerably – although, as Bell says, they did not all seem to get the point: 'The journalists were asking relevant questions like was I taller than or not quite as tall as Princess Margaret (we have quaint units of measurement in Britain) and how many boyfriends did I have at a time?'

The discovery of pulsars led to the awarding of the Nobel Prize for Physics in 1974 to Tony Hewish and his colleague Martin Ryle. Many people, including a number of senior astronomers, were surprised and upset that Bell did not share in the prize and it has remained controversial since. However, although the sexist attitudes of the time may have had some impact on the decision, it was (and still is) very unusual for research students to be recognised by the Nobel committee, something Bell herself agrees with: 'I believe it would demean Nobel Prizes if they were awarded to research students, except in very exceptional cases, and I do not believe this is one of them.' I personally think that she is perhaps being too diffident, as her persistence and hard work were clearly essential to this discovery. As a shy young research student in Glasgow, I was lucky enough to meet her when she took time out of a visit to talk to me and my fellow students about the joys (and frustrations) of a life in science. Her quiet enthusiasm and practical advice were just what I needed at the time to look beyond my initial fears and towards a career in astronomy, and I know many others of my generation feel the same. So, perhaps a more important legacy even than a Nobel Prize is the support, encouragement and inspiration that she has given to many scientists.

In any case, what pulsars show us is that, as well as being very dense, neutron stars can also spin very quickly, some of them rotating several hundred times every second. This is not a surprise, since when things get smaller, they spin faster, but the very high speeds and densities of pulsars mean that they are stretching and pulling the basic fabric of space and time in a way that we cannot hope to recreate on the Earth. Therefore they are very useful as a kind of laboratory for studying basic physics, especially about the nature of 'spacetime' itself and how gravity works.

Eventually, however, all things cease. As pulsars stretch and pull space around them, they slowly lose energy. Young pulsars can spin many thousands of times every second, but as they get older they slow down to maybe one spin every few seconds, and then minutes, hours, days, until eventually they become so slow that they no longer 'pulse' at all and we lose sight of them. They become another invisible neutron star.

Black holes

For even more massive stars, the end result is even more exotic than a neutron star. When the core of a star is more than a few times the mass of the Sun, nothing can stop the collapse – nothing in all of physics – and so it continues and we end up with a bizarre situation in which the star remnant has plenty of mass but no size: it has become a black hole.

Black holes have become part of our culture and one of the most compelling parts of astronomy. Friends, family, strangers on a train – everyone wants me to explain black holes, and the mixture of curiosity and awe, with an undercurrent of fear, that they inspire is something that I have come to immediately recognise. Fortunately, black holes are, at least initially, quite easy to explain and are a simple consequence of extreme gravity.

Standing on the Earth, we are in a moderate gravitational field. If you jump up, you will come back down. If you jump a bit harder, you will go a bit higher and take slightly longer to fall back again, but you will still come back down. However, the faster you jump up, the higher you get, and the longer it takes to return to the ground. If you could somehow jump up with a speed of about 40,000 km per hour, you would be going so fast that the gravity of the Earth would not be quite enough to ever completely slow you down, and so you would never come back down again. That speed is the escape speed (or escape velocity) and anything moving away from the Earth faster than that will never be pulled back.

Of course, if gravity were a bit stronger, the escape velocity would be higher, and you would have to go faster to get away: the escape velocity of Jupiter, for example, is over 160,000 km per hour. The gravity on the surface of an object depends upon two things: its mass and its size. Make the mass larger and the gravity will increase. However, if you keep the mass the same and make the object smaller, the surface gravity will also increase. So, if you imagine crushing the Earth so that it gets smaller and smaller but stays the same mass, the escape velocity on the surface would get higher and higher. When the Earth has been crushed down to the size of the Moon, the escape velocity will be nearly 81,000 km per hour; when it reaches the size of a small town, you would need to go over 1,600,000 km an hour to escape. If the Earth were crushed down to a couple of centimetres across, the escape velocity would become greater than the speed of light.* That means that if you could somehow stand on this crushed Earth and shine a torch straight up, the light would not continue upwards, but would bend over and illuminate your feet. This, of course, would mean that no light can possibly

----

* 1,000,000,000 km per hour or 300,000 km every second.

escape the gravity and so from the outside we would see nothing: the Earth would have become a black hole.

That, then, is the basic concept behind a black hole: an object where the force of gravity is so strong that the escape velocity from the surface is faster than the speed of light. Of course, since nothing can travel faster than the speed of light, not only light but everything else is trapped in a black hole. This gives every black hole a 'point of no return'. From a great distance away there is nothing dangerous about a black hole; if the Sun were to somehow turn into one, it would get very dark and cold, but the Earth would still quite happily orbit around it. But as you get closer to a black hole, the gravitational force gets stronger and stronger until there comes a point where you cannot get away. This distance is called the Schwarzschild radius, and the boundary is the event horizon.

One unfortunate consequence of the one-way nature of black holes is that we cannot tell what is happening beyond the event horizon. Since nothing can get out, no information can escape because there is nothing to carry it, and so the inner workings of black holes are not just unknown but unknowable. This is rather frustrating — and not just for astronomers. One of the most common questions I am asked is what is inside a black hole, but the technically correct answer of 'mass' is clearly unsatis-factory. Because of this frustration, many physicists have tried to speculate about conditions inside the event hori-zon, even though they know their ideas can probably never be confirmed. Extrapolating what we know about the laws of physics outside, these theories all lead to bizarre, even counter-intuitive possibilities: since time slows down near to black holes, inside it is possible that it might even reverse. There is even the possibility of 'wormholes' – where, in effect, two black holes in different parts of the universe join together and provide a 'shortcut' between two different parts of space. Or perhaps a wormhole could con-nect entirely different universes! Unfortunately, exciting though these theoretical ideas are, it is very unlikely that any of them could actually occur in a *real* universe – they require a bit too much fine-tuning and a few too many extremely special circumstances. As a quite pragmatic observational astronomer, I am happy to concentrate on trying to find black holes, and see what they do to things this side of the point of no return.

However, as black holes can't produce any light or emit any particles, even this is far from easy. We can't find them directly, but fortunately we can discover them by looking at the effect they have on things around them.

If some gas crosses the event horizon of a black hole, it is gone from us for ever, making the black hole a bit heavier and the event horizon a bit bigger. However, immediately before it falls into the hole, the gas will be moving very fast and will become churned up and turbulent. This will make it very hot and it will produce a very recognisable signal including X-rays, which we can detect as this is happening outside the black hole. Once the gas has fallen in, the X-rays will stop, but if the gas comes from a star that is in orbit around the black hole – far enough away to be safe, but close enough to lose some of its outer gas layer – then it will not be long before the process repeats itself. These systems which have a black hole and a nearby normal star in orbit around it are called black hole X-ray binaries, and since the discovery of the first by Paul Murdin in 1971, well over a dozen more have been found, although the closest is well over 1,000 light years (9,000,000,000,000,000 km) away. So black holes are all around us – but they are all far enough away not to be scary!

The story of stars brings together many important areas of physics, several of which were unknown until the study of stars provided the clues and measurements needed to develop them. From the extreme gravity and internal forces of black holes and neutron stars, to the creation of the atoms inside us in the cores of stars and supernovae, stars have been at the centre of our expanding knowledge of the laws of nature for centuries. Nevertheless, like most astronomers, when I look up at the night sky, I sometimes temporarily forget that stars are laboratories of cutting-edge physics and simply get lost in the wonder of them twinkling so far above me. And sometimes, if I am lucky and the skies are particularly clear and dark, I can see what I think is the greatest sight the night sky has to offer to the naked eye: countless stars blending together to form the arch of the Milky Way overhead. Which is where we are heading next.

5

GALAXIES

The next time it is clear at night, go out and try to count all the stars you can see. If you are in a town or city with lots of street lights, you will probably only be able to spot a dozen or so, but if it is dark and clear, you will rapidly lose count. From the outskirts of my small home town last night, I lost track after fifty (even though I forgot my glasses), and more careful and patient observers than me in the very darkest places have counted over 500 stars on a single night. However, that is a tiny fraction of the stars out there. With the development of the telescope, not only did many fainter stars become visible, but it became clear that even the vast, beautiful arching 'cloud' of the Milky Way was not a cloud at all, but a mass of countless stars.*

Even now, with photographic surveys and computers to do the work for us, we cannot count all the stars in the Milky Way – many are still too faint to see, or are hidden by clouds of dust and gas. However, we can make a reasonable guess based on what we can see, and it is likely that there are at least 100 billion stars, and perhaps as many as 400 billion. In a world of multi-billionaires and bank bailouts, it is easy to become a bit blasé about millions and billions, but these really are vast numbers. If for some inexplicable reason you did decide to count all those stars, and you went without food or sleep and just kept counting, it would still take several thousand years to count them all.

The bow of the Milky Way across the night sky seems to give us an outside view of our galaxy, but in fact we are well inside it. The whole galaxy of the Milky Way is a giant disc of stars, a bit like two dinner plates put together to keep a dinner warm. We are between the two plates, about a third of the way out from the centre, and the individual stars we see are the ones in the part of the disc immediately around us. What we call the Milky Way

* *

*   See plate section.

– the band of light we see in the sky – is as if we are looking right towards the centre of the plates, where most of the dinner is.

This, of course, should make the Milky Way a vast playground for astronomers. With all of those stars and their attendant planets, together with giant 'nurseries' where new stars are being formed, or the debris of old, dead stars, there is enough to keep anyone busy for several lifetimes. But there is a problem. There may be a lot to explore, but it is all very, very far away.

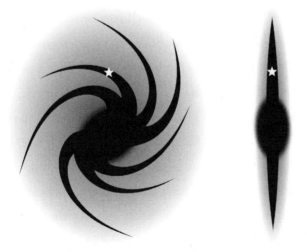

Diagram of the Milky Way. To the left is a plan view, showing the spiral arms radiating out from the central core. To the right is a side view with the thin disk of arms connecting into the central bulge. The approximate position of our Sun is marked with a star on both.

It can be difficult to really appreciate the vast scale of things in the universe. Even the difference between two things we have a real appreciation of the 'largeness' of – such as the Earth and a galaxy like the Milky Way – is hard to grasp. It has become commonplace in the last few decades to comment (even complain) about how 'small' the world has become, but even in the fastest mode of transport any of us are likely to travel in, it still takes around a day to get from one side of the globe to the other.

Consider Concorde, the epitome of rapid transport for a whole generation. Travelling at over 2,000 km each hour, and ignoring the need for fuel stops, it would take Concorde nearly nineteen hours to fly once around the world – a big improvement on the three years it took Magellan's expedition in the sixteenth century. Flying around Jupiter would take around eight days, while the Sun could be circled in less than three months (at least, once the minor problem of stopping the plane from melting had been solved). This is a long time to be stuck on an aircraft, but not beyond the bounds of our imagination. Things become a little more troubling for the very largest stars – 'hypergiants'. With diameters a thousand times that of the Sun, it would take a handful of centuries to circumnavigate those.

But the step up to galaxies is much, much bigger. A supersonic jet like Concorde could fly around the outskirts of a medium-size galaxy like our own Milky Way in around 10^{11} years – that's 100 billion years, or nearly ten times longer than the universe has existed.

The distance between galaxies is an even more mind-boggling number. Astronomers long ago gave up using miles and kilometres to measure distance in the universe and instead starting using the speed of light instead. Light moves very fast – 300,000 km every second. While it would take Concorde about nineteen hours to fly around the world, light could lap you six times before you had even put on your seatbelt. So it's a useful way to measure vast distances. The Sun, for example, is about eight light minutes from the Earth. The Milky Way is around 100,000 light years across. But the nearest galaxy to us – Andromeda – is well over 2 million light years away, and nowadays astronomers regularly look at galaxies billions of light years from us.

It can be hard to get your head around what this really means. At the end of a long, tiring night observing at a telescope in the Canary Islands, while I was waiting for the dome to fully close, I thought about what it was I had actually been looking at all night. The object I was studying was 3C279 – a galaxy with a giant black hole in the centre – and I was working with colleagues to try to work out what was falling into the black hole. However, 3C279 is about 5 billion light years away, and so the light I had spent the night gathering with the telescope had left its galaxy before the Earth had even been formed. Astronomy can really put our lives into perspective.

Of course, it is only quite recently that we have known just how far away other galaxies are, or even what they are. Indeed, with the sorts of distances involved, it's perhaps not surprising that before the invention of the telescope, nothing at all was really known about galaxies. Other than the vast arch of the Milky Way, galaxies had only been visible as a small number of fuzzy 'clouds' which could only be seen on very clear nights. But once astronomers started to sweep the sky with their ever-improving telescopes, they observed many objects that showed a clear structure: not round and solid-looking like the planets, but fuzzy and cloud-like. Collectively called 'nebulae' (from the Latin for 'cloud'), there was initially quite a lot of interest in them, but, perhaps surprisingly, that interest waned quite rapidly. It is not clear exactly why this was. Maybe they were simply lost in the rush of new discoveries, or perhaps it was the challenge of spotting them with the early telescopes.

There was one area that people were very interested in, however, which would later prove immeasurably useful in the study of nebulae, and that was the fashion for comet hunting.

The passing of a comet was thought by many to predict tumultuous events, so once telescopes could be used to spot comets long before they were visible to the unaided eye, people enthusiastically embraced the search. In many ways, finding comets was quite simple, although it required a lot of time and patience. Two things distinguish comets from stars: comets move from night to night, and they are slightly extended and fuzzy. The technique, then, was to choose a small patch of the sky and study it carefully, making a note of anything that looked 'nebulous', and measuring its position. A few nights later, you would return to the same patch of sky and remeasure the position; if it had moved, you had found a comet.

One of the most successful comet hunters was Charles Messier, who had been inspired to study astronomy by a spectacular 'six-tailed comet' he saw as a teenager in 1744. However, Messier became increasingly frustrated that the majority of the objects he found did not move: they were nebulae not comets. To prevent himself from wasting time by returning to them later, and to stop other observers from falling into the same trap, he gradually built up a list of these annoying, fuzzy non-comets which he then published, gradually adding new objects found by him and others until the catalogue reached a total of 103 objects in 1781.*

Messier discovered thirteen comets in twenty-five years – an excellent record. However, it is the catalogue itself that has become his lasting legacy. Although intended as a list of objects to be ignored, it has become the go-to catalogue for interesting and beautiful things to observe through a small telescope, and to this day many of the most interesting objects in the sky are referred to by their Messier Catalogue number (the Orion Nebula, for example, is M42).

However, Messier's catalogue was not systematic, the objects were not described and his telescope was not the best available, so it did not lead to much of an increased interest in nebulae at the time. It took the first great telescopic cataloguer to do that: Sir William Herschel. We have already come across Herschel as the discoverer of

. . . .

* In 1921 and 1966 it was realised that several more objects had actually been observed by Messier or his assistant Pierre Méchain, and so the catalogue was increased to 110.

Uranus (see page 70), but although this is what made him famous, it was his astonishing work 'sweeping' the sky, night after night, with his peerless telescopes to find and record nebulae and star clusters that was his real life's work. Finished after his death by his son John, the New General Catalogue was in many ways the foundation of modern astronomy. Systematic, carefully documented, full of detail and complete, he created a census of the sky that would act as a resource for many years to come. As a result, nebulae again became a hot topic in astronomy.

THE NATURE OF THE NEBULAE

From Messier's catalogue, and certainly from the Herschels', there were two obvious kinds of nebulae: true clouds, and clusters of stars that only looked nebulous if the telescope was not powerful enough. Ignoring the clusters, the clouds themselves could be subdivided: some were irregular and rather like clouds in the daytime sky, others were almost circular and quite neat, while several showed distinct spiral patterns. Over time it became clear that many of the irregular nebulae were clouds of gas, sometimes with new stars forming inside them, while the circular nebulae were the 'planetary nebulae' that we now know are the final stage in the life of a star like the Sun. However, the spiral nebulae remained a mystery right into the early twentieth century.

The first spiral nebula was identified by William Parsons, the third Earl of Rosse. Parsons dedicated his time and wealth – and the labour of his servants – to building larger and larger telescopes, culminating in 1845 with the Leviathan of Parsonstown at his home, Birr Castle. The Leviathan, with its 1.8-metre diameter primary mirror, was the largest telescope in the world for over seventy years. I was lucky enough to be at Birr in 2001 when the newly renovated Leviathan was being put through its paces, and it is an astonishing piece of engineering and optics. It was particularly special to actually look through a large telescope – not something that is normally done with modern professional telescopes, which are designed for cameras and instruments, not the human eye. The view was also significantly improved when, with a single phone call, the current earl had all the local street lights turned off.

Parsons' aim for the Leviathan was to study in detail the nebulae discovered by Messier and the Herschels, and one of the first objects he studied was M51, which showed such a strong spiral shape that it is now known as the Whirlpool. Others were also found, but none could be resolved into individual stars in the way that the star clusters could with a large enough telescope. It wasn't until we had a better understanding of our Milky Way that an interesting connection to the spiral nebulae could be made.

MAPPING THE MILKY WAY

The first serious attempt to map out the Milky Way was again down to William Herschel. He realised that if stars were spread fairly evenly around the Milky Way, then the uneven shape of the Milky Way in the sky – with many more stars in one direction than the other – must mean that we are positioned a bit off-centre. That in turn means that the number of stars we see in each direction will depend upon the overall shape of the Milky Way, and by counting them he could estimate that shape. For example, if the Milky Way were a simple disc, then we would see very few stars looking 'up' out of the plane of the disc, more stars looking along the disc but away from the centre, and the most stars looking towards the centre and out the other side.

Unfortunately, as Herschel himself realised, there are two possible problems with this technique. First, it does not work if stars are not evenly spread around the Milky Way (if they clump towards the centre, for example) and secondly it assumes that it is possible to see all the stars, and that a larger telescope would not see any more. In fact, both of these problems did affect Herschel's result, so his is not an accurate map of the Milky Way, but he did get the basic shape of a disc right.

More recently, we have been able to overcome these problems and create accurate maps of the Milky Way. Modern maps are built by measuring the distance to vast numbers of stars, something that was impossible before telescopes were launched into space. The most recent space telescope to tackle the problem, called GAIA, will map over a billion stars to truly astonishing accuracy. If I could see as well as GAIA, then sitting here in Liverpool, I would be able to measure the width of a human hair in London. With these results we will finally have a detailed map of our home.

With a better understanding of the Milky Way as a disc with a concentration of stars near the centre, people became more interested in a possible connection with the spiral nebulae that they could see. Unfortunately, there was no simple way to tell whether these were separate 'island universes' – like our own Milky Way but very distant – or much smaller gas clouds within the Milky Way itself. The difference is crucial: if they are close to us, then the Milky Way is, in essence, the entire universe with nothing but empty space beyond it; if they are distant, vast collections of stars, then the Milky Way is just one of many galaxies, and the universe could even contain an infinite number of stars. In its own way this is as important a question as the previous argument over whether the Earth is at the centre of the cosmos, and it has similar implications for our place in it.

THE GREAT DEBATE

In the first couple of decades of the twentieth century, the discussion about the true nature of spiral nebulae reached a peak, and in 1920 it was decided to try to settle the question once and for all in a gentlemanly way with the Great Debate, held on 26 April in the Smithsonian Museum of Natural History in Washington. The arguments would be presented by the two great proponents of each side of the debate: Harlow Shapley and Heber Curtis.

Curtis supported the theory of a vast number of other galaxies beyond the Milky Way, and a very large, possibly infinite, universe. He pointed out that the Andromeda nebula appeared to contain other, smaller nebulae, which

would only be possible by a very unlikely chance alignment if it were part of the Milky Way, but would be
entirely expected if it were a galaxy in its own right. He
also pointed out that many spiral nebulae appeared to be
moving very fast relative to us – far faster than any other
kind of nebula, which clearly made them special.

Shapley, on the other hand, believed that the Milky
Way was alone in the universe. If the other nebulae were
separate galaxies, they would have to be at vast distances
from us – many millions of light years – which seemed
impossible. His argument was supported by the work of
Adriaan van Maanen, whose results appeared to show that
one spiral nebula – the Pinwheel – was visibly rotating on
a time scale of years, meaning that if it were as large as the
Milky Way, as Curtis suggested, the outer parts would be
moving at many times the speed of light, which Einstein
had shown was impossible.

The debate hung on van Maanen's measurements, and
eventually it was shown that he had probably not taken
into account small optical defects in his telescope and camera, and that the accuracy he claimed to have seen was not
possible. Now it is universally accepted that not only is
the Milky Way just one of many galaxies, but there are in
fact vast numbers of galaxies in the observable universe.

Indeed, with a large enough telescope, a good camera and plenty of clear nights, it is almost impossible not to find new ones: in one week at a telescope in Hawaii, I helped to discover several thousand galaxies in a patch of sky smaller than the full Moon.

CLASSIFYING GALAXIES

The ubiquity of galaxies does not make them at all dull, though. No matter how many galaxies I observe (or even discover), the impact of their beauty never wanes. The intricacy of so-called 'grand spiral' galaxies, with their swirls of stars and dust, is well known, but to me there is just as much beauty in a faint, distant smudge in the background of an observation that, with patience and careful examination, gradually resolves itself into the simple oval of an elliptical galaxy, or the thin slash of an edge-on spiral galaxy. It does not take much time looking at galaxies to see that these patterns recur and so, once it had been accepted that other galaxies could exist, the next step was to work out what the various types were and to design a suitable classification system.

The first to do this was Edwin Hubble, one of the most important astronomers of the twentieth century, who created the Hubble galaxy classification scheme (also known as the Hubble tuning fork, or the Hubble sequence).

Hubble divided galaxies into five main types: ellipticals, lenticulars, spirals, barred spirals and peculiars. Elliptical galaxies are generally smooth, featureless objects, with visible shapes ranging from circular to more elongated ovals. Lenticular galaxies appear less oval, very stretched and slightly pointed at the narrow ends. Spiral galaxies have a central circular core surrounded by several spiral 'arms' sprouting directly from the core, while in barred spirals the arms emerge from the ends of a 'bar' that joins them to the core. The relatively rare, messy, irregular galaxies are usually grouped together as peculiar galaxies.

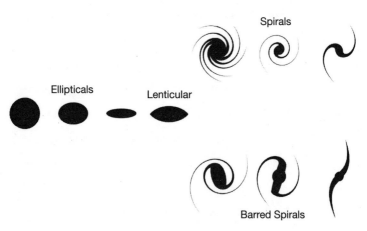

The Hubble tuning fork.

As well as shape, other interesting differences between the groups became apparent, such as colour. Given that the brightest stars are up to a million times brighter than the Sun, the light we see from any group of stars will tend to be dominated by the brightest few, with the majority of stars being too faint to make a difference. So, if we see a blue part of a galaxy, it must mean that the brightest stars there are blue, which means they come from the bright, hot end of the main sequence. As we have seen, such stars do not last long (just a few million years), and so this must be a part of the galaxy where stars are either currently forming, or have done in the recent past: this is a young area. Conversely, in orange and redder regions the brightest stars must be red giants, which appear towards the end of the life of a star, and so these must be old regions.

Elliptical galaxies are almost exclusively orange, whereas the spirals are similarly orange in their core but have much bluer arms and bars. From this, it is clear that only the arms and bars of spirals are home to young stars and recent star formation; ellipticals, lenticulars and the cores of spirals contain only old stars.*

Interestingly, peculiar galaxies tend to be blue, so they

. . . .

* See plate section.

Edwin Hubble

Another one of astronomy's characters, Hubble was born in Missouri in 1889. As well as being an excellent academic student, he was also a gifted athlete. However, following his father's wishes, he chose to ignore his interest in astronomy and science and study law at university, although he kept up with sport, especially boxing – indeed, he was being considered for a world heavyweight title fight when his studies took him from America to Oxford, ending his professional boxing career. However, he never had sufficient motivation to actually practise law, so after a short stint as a teacher, he started a new career as a professional astronomer. In this he was amazingly successful, making some of the core discoveries of the twentieth century, from providing some of the key observations that finally proved the distant nature of galaxies, to discovering the expansion of the universe – the revelation that led directly to the theory of the Big Bang.

must be home to lots of star formation as well. They also often come in pairs, sometimes with streamers of stars and gas flying out from them or connecting the two. Computer simulations have shown that this happens when two galaxies collide or have a very near miss. This happens in part because galaxies can also be gravitationally bound to one another: there are clusters of galaxies.

CLUSTERS OF GALAXIES

Oddly, the first clusters of galaxies were discovered long before galaxies were understood. When producing their catalogues of nebulae, both Messier and Herschel noted that some nebulae seemed to form groups on the sky. In particular, in 1784 Messier noted some nebulae in the constellation of Virgo. When it was confirmed that galaxies are distant objects like our own Milky Way, this was acknowledged as being the first cluster of galaxies – the Virgo Cluster. With his more systematic approach and superior instruments, Herschel identified several others that we now know are clusters of galaxies – in Leo, Ursa Major, Hydra and others.

We now know of many thousands of clusters, ranging in size from a handful of galaxies (usually called a 'group' rather than a full cluster) to hundreds or even thousands of galaxies. They often form their own hierarchies of scale – for example, the Milky Way is part of the Local Group, which in turn is on the outskirts of the Virgo Cluster, forming part of the larger Virgo Supercluster.

The patterns within clusters can tell us a lot about the way they form and change. Rather than containing a simple random scattering of galaxies of all types, larger galaxies tend to be concentrated towards the centre of the cluster, where elliptical galaxies dominate, whereas the galaxies further out are smaller and more likely to be spirals. This is often emphasised by a single, very large elliptical galaxy at – or very near to – the centre of the cluster. These brightest cluster galaxies (BCGs) are the largest galaxies that we know of; some are as much as 100 times the mass of the Milky Way.

The other galaxies in a cluster are not really orbiting the BCG, however; instead everything is orbiting around in a complicated and changing gravitational field created by the mutual interaction of all the galaxies. This means that, over time, as galaxies pass near to each other, they begin to affect each other's orbits – speeding up, slowing down, changing direction. Eventually you end up with galaxies that tend to have long, stretched-out orbits rather than the largely circular orbits of planets in the solar system.

This means that the orbits of galaxies will regularly cross each other. Of course, since galaxies are nearly entirely made up of nothing (think of the vast spaces between the stars within our own galaxy), these 'collisions' are not like two solid objects hitting each other. The galaxies will tend to pass through each other, but this will lead to many interactions between them, as gravity will pull and twist both galaxies, sending a kind of shockwave through them, which can really churn them up. For the stars, this will tend to mean a change in orbit – sometimes from a neat, circular orbit to a more oval one, sometimes even swapping galaxy or being thrown out entirely. The shape of a spiral galaxy is caused by the circular orbits of its stars around the core of the galaxy, so this constant churning will tend to reduce the prominence of the spiral arms; in other words, we would expect this process to gradually turn spiral galaxies into elliptical ones, an evolutionary theory called the Merger Scenario.

This would also explain why galaxies appear as different colours. As we've seen, the formation of new stars requires gas in large clouds, and the shockwaves created during mergers will compress and squeeze previously stable clouds of dust and gas, which is exactly what is needed to start a flurry of new stars forming. However, over time these interactions between the galaxies also tend to strip gas out of galaxies, so eventually there is not enough gas to form new stars – hence, as well as losing their shape, spiral galaxies will also lose the blue areas that indicate the ongoing formation of new stars, leading to red elliptical galaxies which are essentially 'dead'.

Of course, all that gas does not disappear. In the process of being 'stripped' it gets heated up, so instead of collapsing into denser clouds and forming new stars, it spreads out and fills the entire cluster with diffuse, hot gas. It is important to remember that we are talking about something very spread out here. If brought down to the Earth, this gas would appear to be a vacuum, but compared to the emptiness of the space between the clusters of galaxies, it is relatively dense. And very, very hot — typically tens of millions degrees Celsius, which means rather than producing normal visible light like stars, it will produce the more energetic X-rays.

This type of cluster, then, is actually the largest gravitationally bound object that we can 'see'. Not with the naked eye, of course, but using the satellites, telescopes and special cameras that have let us see beyond and explore the full glory of the universe.

QUASARS AND ACTIVE GALAXIES

Another development that helped us to learn more about galaxies started to take place in the 1960s, as we found a new way to explore space: radio astronomy. When they were first proposed, many astronomers were sceptical about building radio telescopes. 'Light' is produced by black-body radiation and although the hottest stars might produce interesting ultraviolet radiation, and cooler ones emit infrared, even the coolest objects could not produce radio waves, so the sceptics asked what would be the point of even looking?

In the 1930s, however, Karl Jansky, an engineer and scientist working for Bell Laboratories, discovered that an annoying source of 'noise' at a frequency of about 20 MHz was being produced by something outside the atmosphere: it was extraterrestrial. Although ignored by many, Jansky's discovery seemed to show that black-body radiation was not the only way 'light' was made, and so it inspired Grote Reber, an avid amateur radio enthusiast, to build a special receiver to try to learn more. He is now credited with building the first radio telescope. His 'radio map' of the sky prompted more interest in the study and, with the detection of radio waves from the Sun in 1942 by James Stanley Hey, radio astronomy began to reach the mainstream. With the construction of enormous radio telescopes like the 76-metre-wide Mk I Telescope at Jodrell Bank, it became possible to produce catalogues of new radio objects.

Not only was this a new way of finding interesting objects to study, it also showed that black-body emission was definitely not the whole story. While stars and nebulae do indeed emit most of their light because they are simply glowing hot, with radio waves we were seeing new ways of making radiation, with electrically charged particles spiralling around in magnetic fields and much more. This was not just new astronomy, but a new insight into the underlying laws of physics.

When a new object is discovered by a radio telescope, we also need to see it with visible light to be certain about what it is – a star, galaxy or nebula that is in the same position. For most radio objects this is a very successful approach. A small number are trickier, but this is probably because the visible object is too faint to be seen; perhaps it is shrouded in dust, which blocks visible light but lets radio waves through. However, in the late 1950s, two objects were discovered that looked like stars, but when examined didn't seem to fit with what we know about stars: either these were made from some exotic new kind of matter (which was not very likely) or something important had been missed.

The objects (called 3C48 and 3C273) were classified as quasi-stellar radio sources or 'quasars', a name that highlighted their mystery. 'Quasi-stellar' in essence means 'they look like stars, but clearly aren't'.

Alongside the puzzle of quasars, other anomalous radio sources started to attract more interest: in particular, a group of fairly normal-looking galaxies with unusually high levels of radio emission, which had been catalogued by Carl Seyfert in the 1940s and then largely ignored. Now all these unusual objects were grouped together as different aspects of the same phenomenon: active galactic nuclei (AGN), or sometimes just active galaxies, which were galaxies with a powerful 'central engine' in their cores that produced vast amounts of energy.

It did not take long to realise that there was only really one candidate for the 'central engine'. When a lot of energy needs to be generated in a relatively small space, lots of gravity is the most efficient way to do it, and the only way to get lots of gravity in a small space is to have a very large black hole. As material falls towards such a black hole, it will speed up and start to collide with other material and produce lots of radiation of all kinds, as well as a strong magnetic field. This will have two important effects: it will produce vast amounts of radio waves and it will 'channel' a lot of the emissions into two narrow 'jets' which are beamed outwards.

If the jet happens to be pointing straight towards us, it will be very bright and swamp out the light from the galaxy, so what we will see will appear to be a small but very luminous 'star' that also produces a lot of radio waves: quasars explained. However, if the jet is pointing off to one side (as it will in most AGN) then we will not get the incredibly bright 'point', but will instead see something that looks rather like a normal galaxy but with unusual radio emission from nearby and perhaps a slightly brighter core than normal: Seyfert's galaxies are also explained.

Astronomers also realised that if a black hole does not have anything falling into it, we will be unable to see it, and so most giant black holes in the centres of galaxies will be invisible most of the time. This means that it is very likely that every single galaxy has one of these supermassive black holes in their cores, but the vast majority just happen to be quiet at the moment.

Of course, if this is true, then our own Milky Way must also have one, and the game was on to try to find other ways to locate it. Fortunately, the gravity of black holes is no different from any other gravity, and so if stars are far enough away from the black hole, they will not be swallowed up and destroyed, but simply orbit it. All we needed to do was see how the stars in the very centre of the

Milky Way were moving around and see if they appeared to be orbiting something very massive but invisible.

Of course, this is far harder than it sounds. The centre of the Milky Way is very crowded, with a confusing mass of millions of stars, and it is also behind a big cloud of dust that blocks much of the light from us. However, in 2008, after a decade of careful observation, and some heroic data analysis, a team of astronomers led by Stefan Gillessen and Reinhard Genzel published the orbits of twenty-eight stars in the very centre of our galaxy. Not only did they show without doubt that there was indeed a black hole there, they were also able to work out its mass: about 3 million times the mass of the Sun.

Since that wonderful discovery, other black holes have been found in other galaxies, with the first image of one released in 2019. The largest known so far weighs in at an astonishing 40 billion solar masses (40,000,000,000 times the mass of the Sun), and it is believed that every galaxy is likely to have such a black hole in its centre. However, it is still uncertain where these black holes come from. It is very likely that they start as small black holes (about 1 solar mass) formed by the death of a large star, and then get bigger as they swallow up stars, gas and even other black holes, but how they become so big is a mystery, particularly since we have found very few 'medium-size' black holes.

GRAVITY AND DARK MATTER

We now have plenty of information and seemingly sensible theories to explain what we can see about galaxies. But as with all aspects of astronomy, new discoveries are constantly being made, and sometimes new information challenges our accepted wisdom, forcing us to rethink our theories. Our observations of galaxies have encountered a mystery that affects our understanding of one of the most fundamental laws: gravity.

The importance of gravity cannot be overstated. All objects owe their existence to this single, ubiquitous force. From the smallest moon to the largest giant elliptical galaxy, all have been created by gravity, collecting, compressing and transforming diffuse clouds of gas and dust.

For nearly 300 years, the work of Kepler and Newton seemed to work perfectly to explain the motions of everything we had observed and discovered in the universe. Based on this, if we look at things orbiting around a galaxy, a very confident prediction would be that they will move more slowly as we look further out.

However, in the late 1950s and through the 60s and 70s, problems started to emerge. In particular, the brilliant observer Vera Rubin found several spiral galaxies that did not rotate quite as expected and, even more importantly, that the speed of orbits around the galaxies did not change, no matter how far out from the centre she looked. More recent measurements show that, for almost all galaxies, the speed of orbiting objects (such as gas clouds or clusters of stars) remains virtually unchanged as far out as we can detect: usually several times further away from the galaxy than the outermost stars.

A related problem came up in the study of clusters. We could now measure all the atoms of the structures we were aware of in the universe, but we encountered a puzzle: there is a discrepancy between the mass of all the atoms we can detect and the total mass implied by the gravitational effects we see.

This all presents a major problem, with only two possible solutions: either we have got gravity wrong in some way, or there is a lot of stuff surrounding galaxies that we cannot see or detect in any way. Both of these suggestions are problematic. If we are wrong about gravity, why does it work so well everywhere else? Perhaps gravity is different in some subtle way in the space between galaxies, or over very large distances, but that seems rather contrived and makes an otherwise very simple and beautiful theory complicated and, to many scientists, rather ugly. However, simply filling the gaps with unseen 'stuff' is also an ugly solution. While it would be a surprise if we could detect all the matter in the universe, this 'missing matter' cannot be any kind of matter that we have ever found before. It cannot be made of atoms, like us and the stars, planets, dust and gas, because then we would be able to detect it. It also cannot be any of the other 'exotic' particles that we have discovered, such as neutrinos, as they would have knock-on effects that we would easily spot. So, not only does there have to be a lot of this unseen substance, but it must be a completely new type of even more exotic matter.

In both these cases of 'missing matter', dark matter may be the solution. This first came to light in 1933. The Swiss astronomer Fritz Zwicky – then working in California – was trying to determine the 'gravitational mass'* of the nearby Coma Cluster by carefully studying the speeds of many galaxies there. Much to his surprise,

. . . .

* The total mass of everything in the cluster, whether it is making light or not, not just the mass of the things we can easily see with telescopes.

the mass he calculated was more than 100 times greater than would be expected just from the stars. He concluded that there must be vast amounts of mass that he could not see, which he called *Dunkle Materie* – the first appearance of dark matter.

There, astonishingly, it rested. Zwicky's findings were largely ignored, and it is only recently that he has started to get the credit he deserves. There are a number of possible reasons for this: at the time, even the idea that galaxies are separate from the Milky Way was fairly new, and anything beyond that was a step too far for many; Zwicky's observations were far in advance of anybody else's and so could not easily be independently tested; one cluster was not really enough for many people to be convinced; and so on. However, probably the most significant reason was much simpler, but less scientific – other astronomers simply did not like Zwicky.

The issue of dark matter is a tricky one. To put this in context, if dark matter does exist, it is not just in the distant reaches of space – it must be everywhere. There will be countless dark matter particles passing through your body every second. However, they do not interact with 'normal' matter and so we do not notice them. In many ways, that is a very good thing (if they collided with normal atoms,

Fritz Zwicky

In many ways, Zwicky was one of the most extraordinary characters of twentieth-century astronomy. He made many fundamental and important discoveries, not just the first evidence for dark matter. He developed the first model for supernovae and neutron stars; he created some of the first truly reliable methods for measuring the distance to remote objects; he produced vast catalogues of galaxies and clusters; and much more. But he got credit for almost none of it in his lifetime. In part, this was simply because he was too far ahead for many of his contemporaries, but his irascible and combative nature certainly did not help.

Although very friendly to support staff and junior researchers, he was brutal to those he considered competitors. He referred to their work as 'the useless trash in the bulging astronomical journals', and in the preface to his superlative catalogue of galaxies he indulges in a rant, insulting other astronomers as 'thieves' and 'fawners' who either stole his ideas or hid their own mistakes and errors.

The flip side to this character was a profound humanitarianism. Appalled by the atom bomb, Zwicky hoped for a future where people would work together without the encumbrance of nations and institutions. He put his ideals into practice as well – after the Second World War, he stockpiled many tonnes of astronomy books and had them sent to devastated areas of Europe and Asia. Even here, though, some of his idiosyncratic side comes through. The collection and distribution of the books was done largely in secret – not due to excess modesty, but because the money to buy the books was not his own, but that of his department at the California Institute of Technology.

In spite of his considerable legacy to astronomy, which is finally getting some of the recognition it has been denied for decades, it is perhaps fitting that Zwicky is most often remembered for his favourite insult: 'Astronomers are spherical bastards. No matter how you look at them, they are just bastards.'

we would all be instantly lethally irradiated!), but it does make them almost impossible to detect, and if we cannot detect them, we cannot work out what they are.

Of course, such a mystery is a gift for lovers of science fiction. Dark matter is a common, if somewhat inconsistent feature in *Star Trek* plots (from dark matter asteroids with bizarre gravitational properties, to aliens made of dark matter); it even forms the inspiration behind Philip

Pullman's very successful *His Dark Materials* trilogy. However, many of these fictional appearances of dark matter miss one of its key properties – its almost non-existent interaction with normal matter means it can have no real effect on our day-to-day lives at all. While this makes it rather dull for science fiction, it is fascinating for science fact – how do you find something that nearly always just passes through any detector we can build?

For decades, astronomers and particle physicists have wrestled with this problem, but with no definitive answers, so the question still remains: are we wrong about gravity, or do we have to find a completely new kind of matter? This is such an important problem: whichever way it goes, we are looking at a fundamental change in our understanding of physics, so it is something we have to continue to explore.

As later work began to confirm Zwicky's results, interest in dark matter increased and astronomers began to look at other advances in the field that might relate to it. In particular, Einstein's theory of general relativity had been confirmed, which explained the effects of gravity on light: gravity bends light, and the more mass there is in an object, the greater the degree of 'bending'. With a simple object like a star, or even a single galaxy, the effect is like a weak lens – a slight focusing of the light from distant objects. However, for clusters, with their complicated structure, the result is much messier – more like badly made sheets of hand-blown glass, all bobbles and lumps. This produces a distorted view of more distant galaxies, revealing arcs, streaks and misshapen blobs which can be used to produce accurate maps of all the mass in a cluster – not just the visible or even X-ray visible matter, but all of it, dark or otherwise. And these maps give the same basic result that Zwicky saw all those years ago: clusters are dominated by a kind of matter that we cannot see or detect and which cannot be atoms or any other kind of particle that we have already found. Dark matter is here to stay.

With all of the important advances that have been made in astronomy and other scientific fields, we can now produce detailed, three-dimensional maps of large parts of the universe around us and catalogues of millions of galaxies, and what we see is rich in structure. There are voids with very few galaxies, dense clusters of galaxies, and giant 'filaments' with galaxies strung out like fairy lights; on these vast scales, the universe looks like a kind of cosmic sponge. These structures show the continuing influence of gravity, and improvements in computer technology have led to precise simulations of the evolution and current structure of the cosmos, with large parts of the universe simulated in detail.

One interesting aspect of galaxies that we have been able to explore is something that Hubble observed: they all seem to be moving away from us, and their speed is increasing with distance. Does that mean we are at the centre of the universe? That seems rather unlikely. Since Galileo and Copernicus, our view of our place in the universe has become one of decreasing 'importance' – the Earth is not the centre of the solar system, the Sun is not the centre of the Milky Way, and the Milky Way is just one of many galaxies. It seems absurd that the Milky Way should somehow be special among all the billions of other galaxies. Another, far more likely explanation is that the whole universe is expanding. Then, no matter which galaxy you happen to live in, all the others will seem to be moving away from you, exactly as Hubble found.

However, if the universe is expanding now, that must mean that it was 'smaller' in the past, and smaller still further back in time, until you reach a time where everything was, somehow, in the same place. If this is true, then the universe had a beginning – and a violent one. The Big Bang.

6

THE BIG BANG

As we have seen, astronomy covers a lot of topics, but the most fundamental questions are about the universe as a whole. Did it have a beginning? If so, when, and what happened? Will it end? There is not a culture anywhere in the world that has not grappled with these questions in one form or another, but until the twentieth century, getting any sort of definitive answer seemed impossible. However, with the discovery of the vast universe outside the Milky Way, and the apparent motion of distant galaxies, a theory emerged that held the promise of conclusive answers even to these fundamental questions: the Big Bang.

The Big Bang theory is now so widely known and well supported that it is easy to forget how revolutionary and radical it was when it was proposed – and how fiercely certain sections of the scientific community resisted it. So how did we arrive at this theory, and what does it actually say?

IN THE BEGINNING

In 1927, when Georges Lemaître first suggested that the mathematical formulae that formed Einstein's general theory of relativity could be used to describe an expanding universe, other scientists were not impressed. Even Einstein himself initially felt that it made no physical sense, although he gradually warmed to the idea.

The theory only really started to be accepted after observations made by Hubble in 1929, which seemed to show galaxies moving away from each other, suggesting that the universe was indeed expanding. But the idea of there being a beginning to the universe – what Lemaître called the 'cosmic egg' – was still difficult for many to accept. How could there possibly be 'a day without yesterday', as Lemaître so graphically put it? Was there a way to have an expanding universe that did not have a beginning and does not evolve?

This was an important point, albeit a largely philosophical one. Essentially, we have to assume that there is nothing special about our place in the universe – something that tends to conflict with many people's view of our world. In the early decades of the twentieth century, many astronomers and physicists felt that the idea of the universe having a beginning – that it could evolve and look different in the past or future – was unacceptable. They thought that there must be an underlying 'perfect' cosmological principle that keeps the universe essentially the same throughout all time *and* space, not just space. In many ways, this was an echo of early thoughts by Plato and others on celestial harmony and heavenly perfection, but was also firmly based in a rational view of the cosmos as something beyond humanity – not something created especially for our benefit, but something that we were an integral but small, largely insignificant part of. Any change in the nature of the cosmos – especially a beginning or end – implied that there was something special about our place in time, if not in space, which seemed to go against our role as simple 'observers'.

An alternative theory was suggested by Fred Hoyle, with his colleagues Hermann Bondi and Tommy Gold. Hoyle said that the idea came to them when they were at the cinema watching a ghost story that ended up exactly where it started, leaving the poor characters in a never-ending cycle of horror. Hoyle later said, 'One tends to think of unchanging situations as being necessarily static. What the ghost-story film did sharply for all three of us was to remove this wrong notion. One can have unchanging situations that are dynamic, as for instance a smoothly flowing river.'

Hoyle's suggestion was that, while galaxies do indeed move apart from each other, the gaps are filled by new galaxies, giving a never-ending, never-beginning universe that changes in detail, but does not evolve overall. His theory was called the 'steady state universe', but there were problems with the idea. In particular, it defied the fundamental law that energy (and hence matter) cannot be created or destroyed (the law of conservation of energy), and many felt that this was a far more difficult problem than a universe that evolved with time.

And so the debate continued, with philosophy dominating over evidence, simply because hard facts concerning the wider universe were rare and often ambiguous. There was a brief pause in further advancement during the Second World War, as all energy tended to be focused on the war effort. After the war, however, the topic was returned to with renewed vigour, and scientists were now also armed with the advances and knowledge gained during the war, particularly in nuclear physics. One of the most important of these scientists was George Gamow.

In spite of, or perhaps because of, a life with more than its fair share of war and revolution, George Gamow was a cheerful man with a well-developed sense of humour and a way with words (his *Mr Tompkins...* books, which explain complicated physics to children, are still popular today). Born in Odessa in 1904, he defected from the Soviet Union in 1933 while at a conference in Brussels. He spent the rest of his life working in the USA, where he became a leading nuclear physicist (although he was not given security clearance to work on the Manhattan Project*).

In the late 1940s, Gamow began to try to answer the apparently simple question: where do all the atoms in the universe come from? Working with his doctoral student Ralph Alpher, he ignored the steady state model and started from an assumption that the universe had a beginning when it must have been very hot and dense – far too hot for atoms to exist in the form we see them today, but

· ·

*   The US-led research and development programme during the Second World War that created the first nuclear weapons.

instead a hot, turbulent mixture of the building blocks of atoms: protons, neutrons and electrons. Gamow called this primeval ocean '*ylem*' (from the Middle English word for 'matter') and he calculated all the different things that would have happened as the *ylem* cooled and expanded.

When the work was ready for publication, Gamow realised that if he inserted another nuclear physicist, Hans Bethe, as an author (although Bethe had not been involved in the work at all), the author list would read 'Alpher, Bethe, Gamow' – too close to the first three letters of the Greek alphabet for him to resist. Unfortunately, he did not tell either Alpher or Bethe before sending the paper for publication. Bethe found out quickly (he was an anonymous editor for the journal) and, since he liked the joke and thought the paper 'had a chance of being correct', he was happy to be included. However, Alpher was less happy, as he felt that much of the credit for perhaps the most important piece of work in his life had been taken from him, and he remained disgruntled up to his death in 2007.

Nevertheless, the '$\alpha\beta\gamma$ paper', as it is now known, was a major breakthrough in our understanding of the universe. The theory was able to predict that the vast majority of the atoms in the universe are either hydrogen (about

75 per cent) or helium (about 24 per cent) — exactly as seen in the spectra of stars. However, its predictions for all the other, heavier kinds of atom were much less successful, producing only a small fraction of those needed. Gamow joked that explaining 99 per cent of the universe should be enough, but it was clearly a problem that the other hundred or so elements in the periodic table were not accounted for.

Hoyle, meanwhile, was looking at the same problem but from a steady state perspective, proposing that all the elements could be made inside stars. In a work that even today is considered one of the most brilliant of the twentieth century, he showed that heavy elements could indeed be made inside stars, and again the amounts matched the observed quantities very well.

With hindsight, it is perhaps obvious that both of these models were right in a way: light hydrogen and helium were created in the early universe, and then processed inside stars over time to make the heavier elements. However, the two models were adopted by opposing sides of the debate on the origin of the universe, which was becoming increasingly heated and even acrimonious.

The now famous name 'Big Bang' arose from the debate, originally suggested by Fred Hoyle as a sarcastic and cheeky description of the theory, but it was taken up by the 'evolutionary' camp with all appearance of gusto! Even the Church got involved, with Pope Pius XII saying that the Big Bang theory was the correct one, since it matched theological thought on an origin of the universe – and Gamow went so far as to suggest that the steady state model was official Communist policy.*

What was needed was hard evidence one way or the other, but that was a huge challenge that was beyond the capabilities of the time, and the advances that were being made were not always reliable at first, often resulting in a return to the drawing board.

For example, it seemed that the steady state theory had scored an early success when Hubble's measured rate of expansion was used to calculate the origin and age of the universe; it produced an age younger than most stars (it was already accepted that stars like the Sun were several billion years old), so clearly something wasn't adding up. However, Walter Baade re-examined Hubble's data and discovered an error in his calculations. Once Baade accounted for this, the recalibrated results showed that the

. . . .

* This was not true: official Soviet thinking was that both theories were 'idealistic' and so unsound.

distant galaxies were actually much further away, and so the age of the universe became comfortably greater than the oldest known stars.

Conversely, the steady state theory predicted that galaxies of different ages would be mixed together, since new ones were forming in the gaps between older ones, while the Big Bang theory would have the oldest galaxies nearby, and progressively younger ones further away (as we look back in time). In 1948, observations by several astronomers seemed to show that more distant galaxies were indeed younger (a point for the Big Bang), but in 1952, Bondi and Gold showed that the data was not really good enough to draw any strong conclusions, so it was back to square one.

By the mid 1950s, opinions were starting to become entrenched. In 1955, Martin Ryle, an early pioneer of radio astronomy, published a large list of newly discovered radio objects that he claimed were a new kind of distant galaxy that was not seen in the nearby universe – a direct contradiction of the steady state theory. However, he rather overstated his case, and Hoyle complained that Ryle was not motivated by a search for truth at all, but simply wanted to discredit the steady state theory (and by extension Hoyle himself).

So things stood for a number of years, with neither

side able to make a significant breakthrough, and with the debate becoming increasingly belligerent.

Then, in 1963, a discovery was published that put the final nail in the steady state coffin. It was a discovery that came from a very unlikely source: the Bell Telephone Company.

Two scientists from Bell Telephone – Arno Penzias and Robert Wilson – were working with early microwave communications technology and trying to characterise the 'noise' in a large microwave receiver. Much to their surprise, they found a noisy signal coming from the sky that was identical no matter where they pointed their antenna.

Their first thought was that there was something wrong with the equipment, and they spent a lot of time carefully examining it. They then wondered if it might be man-made noise, and so pointed the antenna towards New York to see if that made a difference, but the signal remained the same. Finally, they explored the giant 'horn' of the antenna itself and found a quantity of 'white dielectric material' (also known as pigeon droppings). After clearing this out and making sure the pigeons did not return,* they found no

. . . .

* All attempts to scare the pigeons away or prevent them from roosting in the horn failed and so, as Penzias put it, 'To get rid of effect on the signal. It was not man- (or pigeon-) made and it was not coming from the Earth, but from above.

At the same time as this, Robert Dicke, Jim Peebles and David Wilkinson were working in Princeton University – just 64 km away – on a detailed analysis of the Big Bang theory and had actually predicted the existence of a 'glow' of microwaves left over from the early stages of the Big Bang, which would still fill the universe (as there is nowhere else for them to go). They were just in the process of designing an experiment when news of Penzias and Wilson's discovery reached them and they knew they had been scooped.

It is perhaps a shame for Dicke's team that Penzias and Wilson ended up getting almost all the credit (and the Nobel Prize) for their accidental discovery, but it is worth noting that the papers from both groups were deliberately published together in the same edition of *Astrophysical Journal Letters* to avoid any possible conflict, and they continued to share their results and thoughts. It is clear that

. . . .

them, we finally found the most humane thing was to get a shot gun ... and at very close range [we] just killed them instantly. It's not something I'm happy about, but that seemed like the only way out of our dilemma.'

the science was the important thing to everyone involved, not the fame.

The discovery of this cosmic microwave background was a huge step ahead for the Big Bang theory. Not only could it not be explained by the steady state theory (despite various attempts), its discovery was separate from the Big Bang theory itself. It was observed and proven independently.

Although there have been attempts to revive steady state-like models in recent years, the Big Bang is still by far the most successful theory of the origins of the universe and it is now taken for granted by most astronomers (at least until something even better comes along).

So what exactly does the Big Bang theory contain? What is our universe's origin story?

A HISTORY OF THE EARLY UNIVERSE

The key feature of the Big Bang model, as we've seen, is that as the universe has evolved, it has expanded. So if we imagine going back in time, everything moves closer together, making the universe more dense and, importantly, hotter (since the energy that is currently spread out over a large volume was in a much smaller one).

When taken to its logical extreme, this leads to a universe that begins in an unimaginable 'fireball' of heat and density. At the instant of the Big Bang itself, the universe would be almost infinitely dense and hot.

With our current understanding of physics, we cannot hope to explain or even describe those sorts of conditions. However, only a tiny fraction of a second after the Big Bang, the universe would already have cooled a lot, and we can use what we do know to start to understand what may have been going on then – or at least make our best guess!

This stage is called the Planck epoch: just 10^{-42} of a second* after the Big Bang, when the average temperature of the universe was around 10^{32}°C.† At these enormous temperatures, the fundamental forces of nature – gravity, electromagnetism and the nuclear strong and weak forces (which hold atoms together and give them their structure) – cannot be distinguished and so are called 'unified'. At 10^{-32} seconds, the temperature would have dropped by a factor of about a thousand, at which point we believe that gravity, which is by far the weakest of the

. . . .

*   0.000,000,000,000,000,000,000,000,000,000,000,000,000,001 seconds.

†   100,000,000,000,000,000,000,000,000,000,000,000°C.

four forces, would have separated from the other three, and not long afterwards the strong nuclear force would have followed suit.

After about 10^{-12} seconds, the temperature would have dropped to a mere 10^{12}°C — this is the temperature in the middle of the largest particle experiments, such as the Large Hadron Collider at CERN, and so we can start to be a bit more confident about our assumptions as we are able to test our theories about what was going on in the early universe.

Another important feature of this temperature is that it is the highest that we know of at which matter can really exist; when conditions are hotter than this, there is only energy in the form of photons of 'light'. Of course, the particles that existed then were not the ones we are used to — atoms — but exotic particles called quarks and hadrons. In our cool universe, these have stuck together to make the particles we are used to, but in the early universe, the high temperature meant that everything was moving very fast, and the relatively high density of particles resulted in frequent collisions. Therefore any particles that managed to combine together to make something heavier would have been almost instantly ripped apart by smashing into something else.

This continued for a few tens of seconds after the Big Bang, but then the temperature and density both decreased to the point where some of the particles could stick together long enough to form the building blocks of atoms: nuclei made of protons and neutrons. But these were not proper atoms yet – to become normal matter, they needed to have a surrounding cloud of orbiting electrons. Electrons are much smaller and lighter than the nuclei and are not held together by such strong forces, so at this point they were easy to knock away from the central nucleus, meaning atoms could not yet form.

This state of affairs continued without significant changes for nearly 400,000 years. Then, finally, the electrons were able to orbit around the nuclei without getting instantly stripped off again. This is called recombination, and the effect on the universe was enormous. Before this, all the particles in the universe were electrically charged: negatively charged electrons and positively charged atomic nuclei. Since photons of light are electromagnetic in nature, they were strongly affected by the electrically charged particles around them, and got bounced around. This effectively meant that the universe was opaque: light could not travel very far. After recombination, all that changed.

The electrons and nuclei were bound together into atoms, and so the charge of each was cancelled out. Suddenly the photons of light were free to travel in a straight line, and the universe became transparent. At this point the universe was at a temperature of around 3,000°C – about the temperature of an electric fire – and so would have glowed red. However, as the universe expanded and cooled, the light wavelengths shifted beyond the infrared range and towards the radio part of the electromagnetic spectrum.

These are the waves that Penzias and Wilson, the Bell Telephone scientists, discovered – the cosmic microwave background (CMB) – which had provided the closest thing to proof of the Big Bang as anyone could ask for. The importance of the CMB does not stop there, however. It also gives us a direct probe into the very early universe. Prior to recombination, the universe was opaque and so we cannot 'see' anything that happened before then – we have only theories. However, the CMB gives us a 'picture' of the universe at the moment of recombination, just 400,000 years into its 14-billion-year life to date.

At recombination, the entire observable universe was almost all exactly the same: everything was almost the same temperature and density, and so we would expect the

CMB to look the same in all directions and be very 'flat'. However, the word 'almost' is very important here.

Although the universe when the CMB was formed was much smoother than it is now, gravity had already started the process of clumping matter together. The clumps were not very significant, but each contained slightly more matter (and hence mass) than the regions around it, and so each had slightly higher gravity. Photons of light are affected by these patches of gravity, their wavelengths becoming longer as they escape the stronger gravity. Longer wavelengths look cooler (i.e. redder), and so we can detect these changes by measuring the temperature of the CMB in different directions. With very sensitive micro-wave telescopes in space, we can see that some patches can be a bit cooler or hotter. The differences are tiny, but they were first detected by the Cosmic Background Explorer (COBE) satellite in the early 1990s and have since been mapped at higher and higher resolutions.*

It is difficult to convey the importance of these maps. I can distinctly remember, as a postgraduate student in Glasgow in 1992, huddling around a computer screen with most of the rest of the department as the first all-sky

* See plate section.

image from COBE was (very slowly) downloaded. After millennia of astronomers looking at nearby stars and galaxies and trying to understand the cosmos from them, here I was, just starting my career and looking a map of the entire universe as it was about 100 million years before the very first stars even formed. Much as I enjoyed the intellectual challenge of astronomy, I think that was the moment when I realised how exciting it could be.

THE FIRST STARS

The story of the universe since the epoch of recombination has been one that we can recognise in the skies around us. With matter and light free to go their separate ways, gravity really began to make itself felt on the enormous, slightly over-dense clumps that have left their signature on the CMB. After a very long period (probably 100 million years or more), some of those clumps had developed regions in them dense enough for the very first stars to form, with nuclear fusion igniting in their cores: so-called 'first light'. Sometime later the very first galaxies emerged, made from billions of stars, and since then more and more galaxies have formed and the expansion of the universe has continued, pushing them further and further apart.

However, although this story is quite reasonable, some of it is a bit speculative, and there are some gaps in the theory. In particular, exactly when the first stars and galaxies formed is not clear. Unfortunately, there is no simple way of resolving this. The galaxies nearest to us may be old, but stars have been forming and dying in them all through their history, so they don't necessarily contain the earliest stars, and so studying them won't give us the answer. The only way to be certain of when the first stars and galaxies formed is to find very distant galaxies so we are seeing them when the universe was young, and the only stars we see are the very first ones. Of course, this is easier said than done – as we've seen, the more distant an object is, the fainter it looks to us, so to find more distant objects we need bigger telescopes and longer exposures.

There have been a number of very 'deep' observations taken in recent years, particularly by the Hubble Space Telescope. The first of these – the Hubble Deep Field or HDF – took several days of time on the telescope, and contains many thousands of extremely distant (and hence early) previously unknown galaxies, which it was able to show were different from those around us. This was very important as it showed that the universe had changed and evolved significantly since these early galaxies formed, but it did not demonstrate that they were actually the earliest. So, an even longer set of observations was taken: the Hubble Ultra-Deep Field (HUDF), with a total exposure time of about 1 million seconds.* This showed even more galaxies, and again showed significant evolution and change with time, but again it was not clear that these were the earliest. The problem is simply that if you do a deeper observation and find more galaxies, you cannot be sure that an even deeper one will not find yet more. Essentially, the process will only stop when a deeper observation finds nothing new at all.

The HUDF is essentially at the limit of current technology. Longer, deeper observations are possible, but only at the cost of vast amounts of telescope time, which would be much better used for other science. We therefore need to wait for much larger telescopes, which will be able to see further and quicker; then, hopefully, we can find the first stars and so complete our view of the universe.

However, even without that last piece, we can be pretty confident that we have got the story largely right. The first stars formed from the hydrogen and helium created during the latter stages of the Big Bang. Not long after

---

* See plate section.

the first stars came the first supernovae, sending shock-waves and newly formed heavy elements out into the space around them. This triggered new waves of star formation, now with a more complicated mixture of elements, and so things developed until we reached our modern universe with its many stars, galaxies, complicated atoms and molecules and, eventually, us. Dark matter throws a small spanner in the works, but although we may not know exactly what it is, we can work out the effect it has and so include it in our theories and calculations.

The main remaining question, therefore, is what happens next? What will be the ultimate fate of the universe?

THE FUTURE OF THE UNIVERSE

Given the success of the Big Bang model in explaining the past, the next sensible step was to use it to look into the future as well. The theory tells us that the fate of the universe depends upon its average density now. The higher the density, the more mass we have and the stronger gravity will be. Since the Big Bang itself, the universe has been expanding, but gravity wants to pull things together, so it will tend to slow that expansion down.

If there is enough gravity (in other words, a high enough density) then the gravity will one day slow down the expansion completely, halt it and put the universe into reverse. At that point, galaxies will start to get closer together. The closer they get, the stronger the gravity will be and the faster the galaxies will contract, until eventually everything will meet again in a Big Crunch – perhaps creating a single black hole of unimaginable mass, perhaps just disappearing entirely, or perhaps re-exploding in another Big Bang to create another universe. This high-density universe is called a 'closed' universe.

However, if the density is lower than what is required for a closed universe, gravity will never be quite strong enough to stop the expansion, and the galaxies will continue to get further and further apart forever. Eventually all the free gas will either be compressed into dead stars or scattered in the vast voids between galaxies and new star formation will cease. The smallest, oldest stars will eventually 'die' themselves, and the remaining white dwarfs gradually cool down until they are almost at absolute zero. This is the Big Freeze, and a universe of sufficiently low density for this to happen is called 'open'.

However, there is one other option. There is a 'critical' density that marks the border between 'closed' and 'open', and a universe that has exactly that density will also reach a Big Freeze, but much more slowly. This is called a 'flat' universe, and we know that our own universe is surprisingly close to it.

While open, closed and flat universes are all theoretically possible, in practice there are limitations. In particular, if the universe was only a small amount over the critical density in the early stages of the Big Bang, then it would have collapsed in on itself well before the first stars could form. Conversely, if it were even slightly lower than the critical density that early on, the first atoms would have rushed apart so fast that no clouds of gas would ever have been created and again no stars would have formed. So, for the universe around us to exist at all, we must be *very* close to critical density and if we are that close, is it not likely that we are exactly on the balance point? To most people, it seems more reasonable that the universe is exactly flat for a reason we have not yet discovered, as opposed to a nearly-but-not-quite-flat universe just by random chance. There may not be anything in any of our theories yet to say why, but that does not mean that there isn't an underlying reason; we may just not have found it yet.

Measuring the density of the universe was therefore an important thing to do, to work out the future of our universe. But this wasn't easy – we couldn't simply add up everything we could see in the universe around us, mostly because we could never be certain that we hadn't missed something. The safest way, therefore, was to look at how the expansion of the universe has changed with time: if we could measure the current rate of expansion, what it was in the past and how it has changed, that would tell us immediately what kind of universe we live in and how dense it is.

That wasn't possible until the development of large digital cameras and fast computers – identifying far-off galaxies and calculating their distance involved a vast amount of data that needed to be analysed very quickly. However, in the 1990s, two projects were started that had the technology needed: the Supernova Cosmology Project in the USA led by Saul Perlmutter, and the High-Z Supernova Search Team in Australia led by Brian Schmidt and Adam Riess. Both worked independently to study as many distant galaxies as possible and use their observations to see how rapidly the rate of expansion of the universe has slowed down: was it the fast deceleration of a closed universe, the slow deceleration of an open universe, or the perfect balance point between them of a flat universe?

Astonishingly, it turned out to be none of these. Far from decelerating, on the largest scales, the universe was clearly accelerating! This went against any of the predictions. Even in a low-density universe, gravity should still slow the expansion down a bit. The only explanation was that there was an entirely new factor in the universe to consider: something that only acted on very large scales, and so could not be detected nearby, which pushed bits of the universe apart.

As always when something new is discovered, it is important to find a suitable name for it, but this was a completely new idea; no one had any idea what this new phenomenon was and there wasn't really anywhere to start. In the end the 'dark' from dark matter was adopted, and 'energy' was tagged onto it, on the grounds that it must be something like energy if it is accelerating things – and so the term 'dark energy' was coined. However, it is quite likely that when we have a better idea of what dark energy is that the name will become inaccurate and confusing.

But even now, more than a decade after the discovery, we still have no clear idea of what dark energy might be. We do know that it dominates the cosmos: nearly 70 per cent of the content of the universe is dark energy, with dark matter making up most of the rest, and 'normal' matter just 5 per cent. We know that it only works on large scales, so it cannot be something like 'anti-gravity', since that would be noticeable within the solar system. We know that it seems to change with time, although not in a way that has been clearly measured. But we really don't know much more than that; the best theories that anybody has come up with cannot even get close to predicting the effect we actually see.

This is perhaps the most challenging and exciting question in physics, and perhaps all of science, and it emerged just when we thought that we had a theory that accounted for the history and future of the universe. Who knows what discoveries and developments could be made in the next few decades that will challenge everything we think we know? The fun is just beginning.

APPENDIX:
FACTS AND FIGURES

THE WHOLE SKY

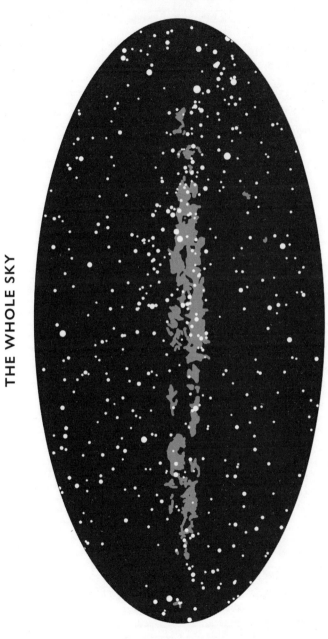

An all-sky map of the brightest 500 stars, with the Milky Way running across the middle.

POWERS OF 10

Decimal	Power of 10	Metric Prefix	Example object (m)
0.000 000 000 000 000 000 000 001	10^{-24}	yocto	Event horizon of a 1 tonne black hole
0.000 000 000 000 000 000 000 01	10^{-23}		
0.000 000 000 000 000 000 000 1	10^{-22}		
0.000 000 000 000 000 000 001	10^{-21}	zepto	
0.000 000 000 000 000 000 01	10^{-20}		
0.000 000 000 000 000 000 1	10^{-19}		
0.000 000 000 000 000 001	10^{-18}	atto	Quark
0.000 000 000 000 000 01	10^{-17}		
0.000 000 000 000 000 1	10^{-16}		
0.000 000 000 000 001	10^{-15}	femto	Proton
0.000 000 000 000 01	10^{-14}		
0.000 000 000 000 1	10^{-13}		
0.000 000 000 001	10^{-12}	pico	
0.000 000 000 01	10^{-11}		Hydrogen atom
0.000 000 000 1	10^{-10}		Water molecule
0.000 000 001	10^{-9}	nano	
0.000 000 01	10^{-8}		
0.000 000 1	10^{-7}		
0.000 001	10^{-6}	micro	Skin cell
0.000 01	10^{-5}		
0.000 1	10^{-4}		
0.001	10^{-3}	milli	Polio virus
0.01	10^{-2}	centi	Marble
0.1	10^{-1}	deci	Potato
1	10^{0}		Human
10	10^{1}	deca	House
100	10^{2}	hecta	Palace
1,000	10^{3}	kilo	Mountain
10,000	10^{4}		Neutron star
100,000	10^{5}		France
1,000,000	10^{6}	mega	Earth
10,000,000	10^{7}		
100,000,000	10^{8}		Jupiter
1,000,000,000	10^{9}	giga	Sun
10,000,000,000	10^{10}		
100,000,000,000	10^{11}		Distance to the Sun
1,000,000,000,000	10^{12}	tera	Largest stars
10,000,000,000,000	10^{13}		Distance to the nearest star
100,000,000,000,000	10^{14}		
1,000,000,000,000,000	10^{15}	peta	
10,000,000,000,000,000	10^{16}		
100,000,000,000,000,000	10^{17}		
1,000,000,000,000,000,000	10^{18}	exa	Distance to Polaris
10,000,000,000,000,000,000	10^{19}		Dwarf Galaxy
100,000,000,000,000,000,000	10^{20}		Distance to the centre of the Milky Way
1,000,000,000,000,000,000,000	10^{21}	zetta	
10,000,000,000,000,000,000,000	10^{22}		Distance to Andromeda
100,000,000,000,000,000,000,000	10^{23}		
1,000,000,000,000,000,000,000,000	10^{24}	yotta	
10,000,000,000,000,000,000,000,000	10^{25}		
100,000,000,000,000,000,000,000,000	10^{26}		Size of observable universe*

* So the observable universe is about 100 yottametres across.

PLANETS OF THE SOLAR SYSTEM

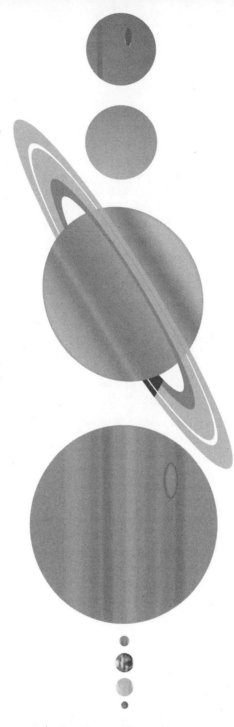

The planets to scale. From left to right: Mercury, Venus, Earth, Mars, Jupiter, Saturn, Uranus and Neptune.

Planet	Type	Size (diameter)	Average distance from Sun	Orbital period
Mercury	Terrestrial	4,880 km	0.38 AU*	88 days
Venus	Terrestrial	12,100 km	0.72 AU	225 days
Earth	Terrestrial	12,760 km	1 AU	1 year
Mars	Terrestrial	6,780 km	1.52 AU	1.88 years
Ceres	Dwarf	946 km	2.77 AU	4.6 years
Jupiter	Gas Giant	139,800 km	5.20 AU	11.9 years
Saturn	Gas Giant	116,500 km	9.45 AU	29.5 years
Uranus	Ice Giant	50,720 km	19.2 AU	84.0 years
Neptune	Ice Giant	49,240 km	30.1 AU	165 years
Pluto	Dwarf	2,380 km	39.5 AU	248 years
Haumea	Dwarf	About 1000 km	43.1 AU	283 years
Makemake	Dwarf	About 1400 km	48.8 AU	310 years
Eris	Dwarf	About 2300 km	67.7 AU	557 years

* AU (or Astronomical Unit) = 149,600,000 km and is the mean distance from the Earth to the Sun

SIZES

Object	Diameter	Mass	
Moon	3,470 km	7.4×10^{22} kg	
Earth	12,760 km	6.0×10^{24} kg	
Jupiter	139,800 km	320 × Earth	
Sun	1,360,000 km	330,000 × Earth	
VY Scuti	1,800 × Sun	10 × Sun	The largest known star
Typical White Dwarf	10,000 km	1 × Sun	
Typical Neutron Star	20 km	1.5 × Sun	
Orion Nebula	24 light years	2000 × Sun	
Large Magellanic Cloud	14,000 light years	10 billion × Sun	The LMC is a small galaxy orbiting around the Milky Way
Milky Way Galaxy	170,000 light years	1000 billion × Sun	
Andromeda Galaxy	200,000 light years	1500 billion × Sun	
IC 1101	400,000 light years	2,500,000 billion × Sun	The largest known galaxy

DISTANCES

Object	Distance	Distance as light-travel-time	Notes
Moon	384,400 km	1.3 light seconds	
Sun	1.5×10^8 km	8 light minutes	
Pluto	5.9×10^9 km	5 light days	
Proxima Centauri	4.0×10^{13} km	4.2 light years	Nearest star to the Solar System
Orion Nebula	1.3×10^{16} km	1,300 light years	
Centre of the Milky Way	2.5×10^{17} km	26,000 light years	
Andromeda	2.4×10^{19} km	2.5 million light years	
Coma Cluster	3.1×10^{21} km	320 million light years	
GRB 090423	1.2×10^{23} km	13 billion light years	Most distant known exploding star
GN-z11	1.3×10^{23} km	13.4 billion light years	Most distant known galaxy

INDEX

Index

ABOUT THE AUTHOR

Andrew Newsam is Professor of Astronomy Education and
Engagement at Liverpool John Moores University. After
studying cosmology at Glasgow University, and working as an observational astronomer at the University of
Southampton, he joined LJMU in 1998 to help set up the
educational arm of the Liverpool Telescope, which later
became the National Schools' Observatory, one of the largest astronomy education projects in the world. As well as
astronomical research and education he is a keen science
communicator, giving talks to many thousands of school-
children, amateur astronomers and the general public
throughout the UK and beyond, as well as working with
artists of all kinds on new ways – from show gardens to
street theatre – to bring the delights of astronomy to as
many people as possible.